Writing Perf

Writing Performances: The Stages of Dorothy L. Sayers

Crystal Downing

palgrave
macmillan

WRITING PERFORMANCES

First published 2004 by
PALGRAVE MACMILLAN™
175 Fifth Avenue, New York, N.Y. 10010 and
Houndmills, Basingstoke, Hampshire, England RG21 6XS
Companies and representatives throughout the world

PALGRAVE MACMILLAN is the global academic imprint of the Palgrave Macmillan division of St. Martin's Press, LLC and of Palgrave Macmillan Ltd. Macmillan® is a registered trademark in the United States, United Kingdom and other countries. Palgrave is a registered trademark in the European Union and other countries.

ISBN 1–4039–6452–1 hardcover

Library of Congress Cataloging-in-Publication Data
Downing, Crystal
Writing performances : the stages of Dorothy L. Sayers / Crystal Downing.
p. cm.
Includes bibliographical references and index.
ISBN 1–4039–6452–1 (alk. Paper)
1. Sayers, Dorothy L. (Dorothy Leigh), 1893–1957—Criticism and interpretation. 2. Detective and mystery stories, English—History and criticism. 3. Women and literature—England—History—20th century. 4. Wimsey, Peter, Lord (Fictitious character) 5. Modernism (Literature)—England. 6. Autobiography in literature. 7. Self in literature. I. Title.

PR6037.A95Z647 2004
813'.52—dc22 2003070738

A catalogue record for this book is available from the British Library.

Design by Newgen Imaging Systems (P) Ltd., Chennai, India.

First edition: August 2004
10 9 8 7 6 5 4 3 2 1

Printed in the United States of America.

For David,
with the mind of a maker

Contents

List of Illustrations

ACKNOWLEDGMENTS

Writing Performances: The Stages of Dorothy L. Sayers was made possible by the institutional support of Messiah College, which provided me with an early sabbatical to write the first draft, a Scholar Chair to finish the project, and a grant to visit Sayers sites in England. The college also funded three student interns to help with final preparations of the manuscript; Rachel Petersen and twins Rebecca and Ruth Brown were invaluable, their careful attention to detail making this a better book. Finally, the Messiah College English Department allowed me to take advantage of the intelligence and good nature of work study student Janel Atlas, who provided both editorial and moral support.

Also crucial to my writing performance has been the Marion E. Wade Center of Wheaton College, which awarded me the Clyde S. Kilby Research Grant for 2001. The staff at the Wade have been gracious and supportive, not only allowing me to quote unpublished Sayers documents I discovered in their reading room, but also making available the photographs of Sayers that grace this book.

There are many specific individuals who have helped shape the thought that has gone into *Writing Performances*, but two in particular deserve special attention. First is Barbara Luton, who gave me *Whose Body?*, my first Lord Peter Wimsey novel, and then insisted that I get my own body into graduate school. Next is my most astute reader and best friend, David C. Downing, whose trips to the Wade Center for research on C. S. Lewis provided me with the motivation to study C. S. Lewis's friend.

Janice Brown, author of *The Seven Deadly Sins in the Work of Dorothy L. Sayers*, deserves accolades for a Sayers conference, replete with a full production of *Busman's Honeymoon*, that she organized on the campus of Grove City College in Pennsylvania (October 2000). There I met Laura Simmons, who sent me a copy of Sayers's essay "Sacred Plays," and whose forthcoming book on Sayers's theology I look forward to reading. At the conference I was especially enriched by the company and conversation of Flora Armetta, who, I suspect, will write her own book on Sayers someday.

I also wish to thank my colleagues at Messiah College, who have contributed to an intellectually stimulating and personally encouraging environment. Two in particular have left their mark on this book: theater historian Valerie Smith, who introduced me to performance studies, and church historian Douglas Jacobsen, who introduced me to Radical Orthodoxy and other important constructs energizing my performance. No less important are sculptor Ted Prescott and painter Cathy Prescott, whose joy and insight during numerous lengthy conversations about creativity, theology, sometimes Sayers, more often C. S. Lewis, and too often postmodernism, infused this project with love.

Permissions

Permission has been granted by David Higham Associates to quote various extracts from the poems, plays, and prose of Dorothy L. Sayers (Sayers/05.08.03).

Permission has been granted by The Marion E. Wade Center at Wheaton College to quote from their holdings of unpublished letters by Dorothy L. Sayers.

Permission has been granted to quote Mikhail Bakhtin from: *Art and Answerability: Early Philosophical Essays* by M. M. Bakhtin, edited by Michael Holquist and Vadim Liapunov, translation and notes by Vadim Liapunov, Copyright © 1990. Courtesy of the University of Texas Press; *The Dialogic Imagination: Four Essays* by M. M. Bakhtin, edited by Michael Holquist, translated by Caryl Emerson and Michael Holquist, Copyright © 1981. Courtesy of the University of Texas Press; *Toward a Philosophy of the Act* by M. M. Bakhtin, translated by Vadim Liapunov, edited by Vadim Liapunov and Michael Holquist, Copyright © 1993. Courtesy of the University of Texas Press.

Abbreviations for Frequently Quoted Works by Sayers, with Original Publication Date

Cat	*Cat o' Mary* (unfinished fictional autobiography)	2002
Creed	*Creed or Chaos?* (essays)	1949
Death	*Unnatural Death* (Lord Peter Wimsey novel)	1927
Doc	*The Documents in the Case* (a detective novel)	1930
EC	"My Edwardian Childhood" (unfinished memoir)	2002
Further	*Futher Papers on Dante* (essays)	1957
Gaudy	*Gaudy Night* (Lord Peter Wimsey novel)	1935
GN	"Gaudy Night" (essay)	1937
Ltrs 1	*The Letters of Dorothy L. Sayers.* Vol. 1	1995
Ltrs 2	*The Letters of Dorothy L. Sayers.* Vol. 2	1997
Ltrs 3	*The Letters of Dorothy L. Sayers.* Vol. 3	1998
Ltrs 4	*The Letters of Dorothy L. Sayers.* Vol. 4	2000
King	*The Man Born to Be King* (radio play cycle)	1943
Mind	*The Mind of the Maker* (literary theory)	1941
Search	*The Poetry of Search and the Poetry of Statement* (essays)	1963
Whimsical	*The Whimsical Christian* (essays)	1978
Zeal	*The Zeal of Thy House* (play)	1937

Setting the Stage: An Introduction

. . . every one of us is on the stage,
performing a part in a play,
of which we have not seen either the script
or any synopsis of the ensuing acts.

—Sayers, *The Mind of the Maker*

When Lord Peter Wimsey, dressed as a harlequin, dives from atop a tall fountain into the shallow fishpond below, awestruck (albeit tipsy) partygoers breathlessly applaud his death-defying act. Later he offers a terse explanation for his flamboyant performance: "It pays to advertise." In this scene from her 1933 novel, *Murder Must Advertise*, Dorothy L. Sayers advertises her own love of performance. From ornately costumed theatricals in her parents' home to plays put up at Oxford University, Sayers relished role-playing. Even in her professional life she manifested a flair for the histrionic, with showy baubles dangling from her ears and, occasionally, a Marie Antoinette wig covering her hair loss. It is as though Sayers recognized that the actions one performs become a type of advertisement for the self—explaining, perhaps, why she hid her baby conceived outside of marriage; it was not a performance she wanted to advertise. Nevertheless, when she discussed her son with the cousin secretly raising him, she wrote, "I'm glad he's such a good advertisement. I always knew he was going to be frightfully tough and obstreperous" (*Ltrs* 1. 216). Sayers employs two adjectives that might well describe herself, displayed when she sends B.B.C. representatives not a note of dismay for changes made in one of her radio plays, but an envelope filled with tiny, torn-up bits of her contract. As she states in the opening paragraph of "My Edwardian Childhood," an autobiography that she later abandoned, "I never could resist a touch of the dramatic" (3).

Sayers could not resist giving this touch of the dramatic to Lord Peter, who often plays a role, advertising a self he wants the public to see. In *Murder Must Advertise*, for example, Peter dismisses the fountain-dive as "vulgar ostentation" on the part of his cousin, Death Bredon, who keeps "impersonating" him (200–01). Ironically, this statement is itself a performance: Peter is actually impersonating the fictitious Mr. Bredon in order to investigate a murder at an advertising agency. Dressed as a harlequin, Peter puts on an act of someone putting on an act. When Scotland Yard asks him to maintain the performance of both characters, Bredon and Wimsey, Peter states, "I will advertise myself in some way. Advertising is my long suit" (328).[1]

Though Peter employs impersonations to determine the identity of criminals, in "My Edwardian Childhood" Sayers implies that impersonations are determinate of identity itself: "I . . . have at all periods of my life continued to dramatise myself, into a great number of egotistical impersonations of a very common type, making myself the heroine . . . of countless dramatic situations" (11). In *Cat o' Mary*, an unfinished fictionalized expansion of the earlier memoir, family members discuss the identity of Katherine Lammas, a thinly veiled portrait of Dorothy. When one aunt complains that Katherine "is a mass of affectations," Great Aunt Agatha responds,

> Shows her brains. She's choosing a character for herself. . . . You don't know what 'natural' means. Nor do I. Nor does anybody. Man's an unnatural animal. Makes himself in his own image. The child's always acting—quite right too. She'll have to try herself out in a lot of characters before she finds the one that suits her. (63)

Taking a cue from Great Aunt Agatha—who implies that identity is performative—this study will explore the multiple characters enacted by Dorothy L. Sayers: performances staged through her writing. For Sayers, writing performances—both those she read and those she wrote—seemed more real than her quotidian self. When she transposed her own *Three Musketeers* "dressing-up enthusiasm" into the psyche of Katherine Lamas, she wrote that "the character of Athos was more real to her than her own." Significantly, the performance of Athos resulted in writing performances: "she started to write cavalier poems with swaggering choruses" (*Cat* 59)—similar, perhaps, to the hyperbolic proclamations of love that the fourteen-year-old Dorothy wrote, in the voice of Athos, to her cousin Ivy (*Ltrs* 1. 5–13). Later in *Cat o' Mary*, when Katherine "wrote the words of a Cantata on the Death of Athos, a Ballad on the Death of d'Artagnan, a Farewell Ode to Porthos" as she faced leaving for boarding school, she felt that the world of the Three Musketeers was "all far more real than French lessons and . . . long summer days beneath the mulberry tree" (77–78). For Katherine and her author, writing performances superseded performances of the body.

True to Great Aunt Agatha's predictions about Katherine's multiple "characters," Sayers staged her authorial self in different, albeit overlapping, stages: from poet and editorial assistant to detective fiction novelist and advertising copy writer; then from playwright and literary critic to Christian apologist and Dante scholar. She tried herself out in a lot of authorial characters before she found the one that suited her best, ending her life consumed by the translation and annotation of Dante's *Divine Comedy*. Furthermore, similar to Peter, who convinces newspaper audiences that Lord Peter Wimsey and Death Bredon, who look exactly alike, are "real" persons, Sayers has convinced diverse audiences that her "real" character is manifest in only one particular writing performance. Fans of Lord Peter Wimsey are surprised to discover that Sayers became an outspoken Christian apologist; theologians who cite her work on the Trinity are seldom aware that Sayers first explored the doctrine as a playwright; educators who participate in the pedagogical system inspired by her essay "The Lost Tools of Learning" are amazed that she wrote detective fiction.[2] As Sayers herself once put it, "it has greatly disconcerted the critics who seem to think it quite indecent that a detective story writer would deal in religion! I am reminded of your clerihew about Belloc:

> He seems to think that nobody minds
> His books being all of different kinds." (*Ltrs* 2. 33)

Writing Performances: The Stages of Dorothy L. Sayers discusses the "different kinds" of genres Sayers performed, including those produced on literal stages; however, it does so in order to focus upon writing itself as performance: the performance of cultural values, of the dogmas to which one has given assent, of the language in and through which meaning is inscribed.

Since the 1960s, the concept of "performance" has broadened to include "popular entertainments, speech acts, folklore, political demonstrations, conference behaviour, rituals, medical and religious healing, and aspects of everyday life."[3] Andrew Parker and Eve Kosofsky Sedgwick, in their introduction to *Performativity and Performance*, suggest that "philosophical essays themselves surely count as . . . performative"; and Carlos Alonso recognizes his own "critical act as a performance, an activity that occurs at a given place and time and is dependent on . . . a theater of representation to give it context and authority."[4] For these critics, the *act* of writing performs various cultural scripts—not only about language usage and genre expectations, but also according to the social codes, identity constructions (race, gender, creed, sexuality) and philosophical assumptions that underwrite culture. It is in this spirit that *Writing Performances* examines the theaters of representation that staged the diverse authorial activities *both of and about* Dorothy L. Sayers, assessing the contexts which give authority to their performance.

Ironically, the performances that brought Sayers fame were those undertaken primarily for money. *Whose Body?*, the first Lord Peter Wimsey novel, was begun in 1921 to stave off the wolf at the door—or at least the father in the rectory, who still had to support his twenty-seven-year-old daughter. Before the novel was published, however, Sayers found a job as a copywriter for S. H. Benson, a London advertising firm that was to inspire *Murder Must Advertise*, written after she had finished her nine-year stint in the corporate world.

More than any other of her twelve completed detective novels, *Murder Must Advertise* spectacularizes the performance of writing. Sayers fills her ninth work with histrionic advertising slogans: a margarine ad parodies Dickens with the proclamation "It's a far, far butter thing"; a nerve remedy cites Tennyson's "Tears, Idle Tears"; Bunbury's Whole-Meal Flour is said to take "the Ache out of Baking" (26, 237, 34). Furthermore, Sayers's fictional copywriters at "Pym's Publicity" are so immersed in the language of sales that they often interpret personal experience according to how advertising performs. Miss Meteyard subverts an argument when she realizes that the phrase she just uttered, "Tell that to the marines," would make "a good headline for something-or-other" (182). Mr. Copley's satisfaction over a job well done takes the form of ad copy—"The words sang themselves into a little slogan in [his] head: 'You can Count on Copley in a Crisis' "—thus proving what the narrator told us earlier: that "long advertising practice had given him a trick of thinking in capital letters" (134, 132).

Even Lord Peter, undercover as a copywriter at Pym's, starts performing the language of advertising. Responding to teatime banter about the rigors of copy layout, he intones "If the lay-out lays you out, rejuvenate your soul in Roof Revels with Copy-writers" (94). Later, when accusing Inspector Charles Parker of slow-wittedness about a suspect, he states, "Are you tired at the end of the day? Do you suffer from torpor and lethargy after meals? Try Sparkletone, the invigorating vegetable saline that stimulates while it cleanses" (274). Like Wimsey, Pym's employees perform their copywriting so well that advertising starts to perform them, scripting their thoughts and directing their actions.

This double sense of "writing performances," wherein the self is both in control of and subject to language, directs my study of Sayers, who, I will argue, both reflects and deflects the "slogans" of her day. Like advertisements, different vocabularies in a culture compete for attention, providing divergent, often conflicting, answers to human need. Sayers, like all of us, embraced some vocabularies while eschewing others as insufficient reflections of "truth." What singles her out from many consumers of culture, however, is her realization that all vocabularies that seek to capture "truth" are historically situated constructions that must be questioned and challenged.

Recognizing that no language is innocent, including the hegemonic vocabularies that defined her era, Sayers, this book will argue, anticipates the postmodern ironizing of Enlightenment humanism and scientific objectivity. She would most certainly agree with postmodern philosopher Richard Rorty that "the Enlightenment wove much of its political rhetoric around a picture of the scientist as a sort of priest, someone who achieved contact with nonhuman truth by being 'logical,' 'methodical,' and 'objective.'"[5] While the words "logical" and "objective" seem innocent, during the height of logical positivism in the 1920s and 1930s, they conveyed contempt for any kind of language that referenced belief in unseen realities.

In response to the modernist disdain for religious vocabularies, Sayers makes the case, as do most postmodern theorists, that the modernist absolutizing of "reason" is merely a substitutionary faith—that reason itself is not innocent. Significantly, the mystery Sayers creates for *Murder Must Advertise* is ground in language that is *literally* not innocent. The copy written for Nutrax, a product that settles the nerves, is inscribed with another agenda: each week it signals a new cocaine distribution site. Like Peter Wimsey, the ad copy performs two roles at once. When Peter shares this discovery with his brother-in-law, Inspector Parker states "As far as I can make out, all advertisers are dope-merchants" (251), implying that the Nutrax ads, like dope, are mind-altering.

Many critical theorists today assert that all performances of language are mind-altering. For them, human perception is regulated by "discourse": the patterns embedded in language that delimit what can and cannot be thought. For example, in early modern England it was nearly impossible to think of a tan woman as beautiful because the word "tan" was part of a socioeconomic system that identified bronzed skin as distinctive of the peasant class which had to work outdoors. By the mid-twentieth century, the word "tan," though retaining the denotation of sun-browned skin, had developed an entirely different connotation because of socioeconomic changes; a tan signaled the leisure to be outdoors, to vacation in sunny climates, to join a country club. There is no "logical" or "objective" standard for skin-beauty; it is embedded in discourse about wealth and status that directs human behavior.

Poststructuralists assume that all performances—whether sunbathing or novel-writing—are "always already" written. In other words, language writes our thoughts. Jacques Derrida, one of the parents of poststructuralism, has shown that when Plato sought to establish speech as more authentic than writing because it is closer to the origin of meaning—a self which voices the truths it perceives—Plato couldn't keep out of his philosophic system the performance of "writing," seen when he states in the *Phaedrus* that speech is *written* in the soul of the listener. Thus, for Derrida, even voiced language is a writing performance: "There is nothing outside of the text."[6] For most poststructuralists, then, identity itself is defined by discourse, such that one's sense of self is merely a result of the language one performs; language, in effect, performs us. Similar to Roland Barthes, who proclaims that "only language acts, 'performs,' and not 'me,' "[7] Judith Butler asserts that identity is not self-consciously performed in a type of "theatrical self-presentation"; the self is instead constituted by the performativity of language which regularizes human action:

> [P]erformativity cannot be understood outside of a process of iterability, a regularized and constrained repetition of norms. . . . This iterability implies that "performance" is not a singular "act" or event, but a ritualized production, a ritual reiterated under and through constraint, under and through the force of prohibition and taboo, with the threat of ostracism and even death controlling and compelling the shape of the production.[8]

This assumption, of course, is itself not innocent, and has been questioned by theorists concerned about human agency, about the ability to choose performances that might remediate injustices in the world.[9]

Sayers highly valued choice, both its freedom and responsibilities, and, therefore, just as she demurred from an entirely deterministic view of divine sovereignty,[10] she most certainly would have balked at a deterministic view of language—what Richard Rorty calls the "apotheosis of language," wherein discourse becomes "a kind of divinity, something of which human beings are mere emanations" (*CIS* 11). In *Contingency, Irony, and Solidarity,* Rorty presents a paradox of free will and determinism, wherein a society's perceptions and practices are regulated by the language its members speak—a vocabulary that is entirely arbitrary, without any grounding in or direct access to absolute truth. Certain individuals, however, recognizing the contingency of their society's vocabulary, are able to ironize it, drawing attention to its insufficiencies and its regulatory powers by creating metaphors that capture human experience in new ways. Rorty celebrates the minds of such makers, calling them "strong poets."

Writing Performances demonstrates how Sayers performed the role of Rorty's "strong poet"—with a big difference.[11] While Rorty has given up on religious vocabularies as no longer useful, believing that "the nature of God" is "an unprofitable topic" (*CIS* 8), Sayers would assert that the ironizing of religious vocabularies is necessary precisely because the nature of God is meant to be known but language inevitably gets in our way. She, like Rorty and other postmodern theorists, as we will see, must fall back on paradox to argue for human agency; however, for Sayers, paradox is itself a reflection of the numinous: of that which cannot be contained in language. As Wittgenstein once put it, "There are, indeed, things that cannot be put into words. They make themselves manifest. They are what is mystical."[12]

Sayers establishes a mystical moment in *Murder Must Advertise* as Lord Peter changes in a taxi from his Pym's identity back into his monocled aristocratic persona: "With a vacant wonder he gazed upon the twinkling sky-signs, as though, ignorant astronomer, he knew nothing of the creative hands that had set these lesser lights to rule the night" (107). Sayers's ambiguous "as though" leaves us guessing: is Peter's vacant response to the sky-signs itself a sign that he is "ignorant" of their creator? Or is his gesture yet another performance? Just as he acts out the role of Death Bredon, is Peter merely acting "as though" he knows nothing about the writer of the sky-signs?

Significantly, the word "sky-signs" in Sayers's day referred not only to stellar phenomena but also to advertisements "so constructed and placed that the letters, etc., stand out against the sky" (*OED*). Sayers signals this double meaning of "sky-signs" with the word "astronomer," subtly paralleling the creation of advertisements with the creation of the universe.[13] This parallel gives added weight to an earlier conversation among the Pym copywriters. Recounting the "intimate questions" advertisements ask—"Do you Suffer from Superfluous Hair? . . . Do you ever ask yourself about Body-Odour?"—Ingleby wonders "why the long-suffering public doesn't rise up and slay us." Another copywriter responds by focusing on origins: "They don't know of our existence.... They all think advertisements write themselves" (65). Sayers believed that, just as some advertisements are sky-signs created by hands adept at writing performances, so stellar sky-signs advertise the "creative hands" that formed the universe—whether a society wants to acknowledge those hands or not.

The longer she lived, the more emphatically Sayers discussed creative hands—both human and divine—while also recognizing that Christian advertisements of God often murdered the message through badly scripted copy. Committed to Anglo-Catholicism at a time when modernist attacks on religion were generating fundamentalist backlashes, Sayers developed new vocabularies to describe the "judicial murder of God" (*Ltrs* 4. 139), believing, indeed, that murder must advertise. As she writes in "The Dogma is the Drama,"

> Let us, in Heaven's name, drag out the Divine Drama from under the dreadful accumulation of slipshod thinking and trashy sentiment heaped upon it, and set it on an open stage to startle the world into some sort of vigorous reaction. If the pious are the first to be shocked, so much the worse for the pious.... (*Creed* 24)

This "open stage" was one of many upon which Sayers startled the world. Because she did not stage her writing performances according to the privileged scripts of her time, Sayers garnered, in the various stages of her career, disdain from belletrists for her mystery novels, dismay from her fans when she dismissed detective fiction, horror from fundamentalists for the sacrilegious language of her plays, contempt from positivists for her Christian apologetics, and disgust from orthodox Dante scholars for her translation of *The Divine Comedy.*

Despite Sayers's ability to recognize and subvert the hegemonic vocabularies of her day, scholars trained in critical theory have shown little interest in her, and students of Sayers have failed to address metacritical issues, ignoring the situatedness of her theories—performed by an Anglo-Catholic at the height of modernism—as well as the positionality of theories about her. Furthermore, while her friend and fellow Anglican C. S. Lewis has generated over 150 biographical and critical books, full-length studies of Sayers could be counted on one's fingers and toes. Worse, as Catherine Kenney notes, "It is a great irony that a writer who repeatedly asserted the limitations of the biographical approach to literature has been the subject mainly of biographical studies."[14] Each new biography on Sayers signals the desire for a script that more "accurately" stages the performance of the author's life.

In Sayers's case, the need for greater accuracy is somewhat warranted. In 1975, almost twenty years after Sayers's death, Janet Hitchman revealed to the world that Sayers had given birth to the boy that she unofficially "adopted" when he was eleven years old—even though she never invited him into her home. In 1979, Ralph Hone speculated about the father of the child—incorrectly, as illuminated in James Brabazon's 1981 biography, authorized by Sayers's son. The identity of the father, however, was not revealed until Barbara Reynolds' 1993 centenary biography. This list may explain why Sayers repudiated biographical criticism; with each new revelation she has been reduced to her sexual indiscretion rather than expanded to fill the multiple stages of her subversive perceptivity. This is not to say that biographers have failed to explore and celebrate Sayers's insight; it's just that, in the process, they often perform their own assumptions as they imbue biographical facts with meaning. Biography, as we will see, is a type of writing performance.

The dearth of theoretically informed scholarship on Sayers may result from the way she was largely overlooked, if not attacked, by literary critics of her own day. Sayers began writing her first Peter Wimsey novel the same year the Newbolt Report on "The Teaching of English in England" sounded a religious call: "literature is not just a subject for academic study, but one of the chief temples of the human spirit, in which all should worship."[15] Those disciples who followed this prophetic call relegated detective fiction to an outhouse behind the temple; it was not "Capital A" Art, as a one-time friend of Sayers put it. In fact, literary critics felt responsible to protect the temple of "Capital L" Literature from invasions by mass culture. Their Solomon was Cambridge don I. A. Richards, who believed that poetry "is capable of saving us; it is a perfectly possible means of overcoming chaos." Following in his footsteps were F. R. Leavis and Q. D. Leavis, who railed against the mind-numbing power of advertising, bestsellers, and radio broadcasting—three stages upon which Sayers excelled.[16]

Academic contempt for these types of writing might explain why the three volume *No Man's Land: The Place of the Woman Writer in the Twentieth Century* relegates Sayers to single sentences only twice in twelve hundred pages—one time, significantly enough, as an example of "theatrical transvestism."[17] Nearly exiling her from the "no detective writer's land" of criticism, the famous authors, Sandra M. Gilbert and Susan Gubar, demonstrate the problem with contemporary Sayers scholarship: because Sayers is identified with detective fiction, people suspicious of the literary credentials and/or conservativism of the genre dismiss her, while those who respect detective fiction focus only upon her mystery narratives—the writing performances which Sayers valued least.[18]

However, because postmodern thinkers have undermined the modernist divide between capital "C" and lowercase "culture," detective fiction as well as religious discourse (not to mention advertising, bestsellers, and radio broadcasting) have become fruitful topics for scholarly inquiry. As a result, Sayers's name is reappearing in contemporary criticism. Shakespeare scholar and cultural critic Marjorie Garber mentions Peter Wimsey in *Academic Instincts* (2001) as she discusses distinctions between amateurs and professionals. More substantively, Jacques Barzun's monumental cultural history, *From Dawn to Decadence: 1500 to the Present* (2000), gives Sayers her own heading and three pages of discussion. In contrast, he only briefly mentions her contemporaries Gertrude Stein, Ezra Pound, E. M. Forster, and Agatha Christie (twice each), I. A. Richards once, and the Leavises not at all. Barzun's work reflects a trend in the criticism of the late twentieth century: a reassessment of the modernist canon, and especially woman's place within it. For Twayne's Women and Literature Series, Mary Loeffelholz published in 1992 *Experimental Lives: Women & Literature 1900–1945*, devoting a chapter to Sayers's "vision of social transformation" in her detective fiction. Sayers also gets a chapter in Gill Plain's 1996 study, *Women's Fiction of the Second World War: Gender, Power and Resistance.*

Italian scholar Barbara Reynolds, who finished translating *The Divine Comedy* for Penguin after Sayers died, is responsible for the best Sayers scholarship to date. Not only did she produce *The Passionate Intellect* (1989), an unsurpassable work on Sayers's engagement with Dante, she also managed a Herculean task: editing and annotating four volumes of Sayers's correspondence as well as "My Edwardian Childhood" and *Cat o' Mary.* Reynolds' notes and explanatory contexts make these volumes the best means by which to get a sense of Sayers's beliefs, affections, and practices.

Writing Performances assesses how Sayers grounds those beliefs, affections, and practices, demonstrating the ways she mimics and critiques the discourses that shaped her sensibilities. Her constructs are often paralleled with those of literary critic and moral philosopher Mikhail Bakhtin. Born in Russia in 1895, two years after Sayers's birth, Bakhtin was exiled in 1929 for Christian affiliations that made him sensitive, like Sayers, to the limitations of modernist discourse.

However, unlike Sayers, whose outspoken advocacy of Christian dogma rendered in popularistic terms makes members of the academy uncomfortable, Bakhtin has been celebrated by scholars in many different fields—feminist theory, film, literary criticism, cultural studies, ethics—perhaps because his religious assumptions were suppressed by Soviet totalitarianism. Though Sayers and Bakhtin performed in different theaters of representation, resonances between their writings support the primary argument of this book: that the writing performances of Sayers, many of which adumbrate postmodern paradigms, are more theoretically sophisticated than has yet been acknowledged.

The first half of *Writing Performances* looks at the interpenetration between Sayers's life performances and her writing performances. Chapter 1 presents Sayers's struggle with autobiography as evidence of her suspicions about language, of her recognition that staging "the self" is inevitably a writing performance. Nevertheless, as this chapter will show, biographers rely heavily upon Sayers's artificial stagings of her childhood, and thus end up performing their own values as they seek to illuminate the identity of Sayers.

Chapter 2 focuses on another aspect of performing "the self"—gender—discussing how Sayers both mirrors and subverts cultural constructions of gender not only in her writing performances, but also through the performances of her body. While most critics align Sayers with Harriet Vane, the love-interest she created halfway through her Peter Wimsey canon, Harriet is here presented as the source of dialogic tension, performing a script that is just one among many subject positions Sayers sensed within herself. These multiple positions explain why Sayers wrote explicitly feminist essays while simultaneously feeling uncomfortable with the totalizing discourses of feminist ideology.

Chapter 3 provides a different approach to the relation between life and writing performances, consonant, I believe, with Sayers's negative attitude toward biography. She strongly believed that a writing performance should draw attention, not to the author, but to the work itself, whether the work was a detective novel or church dogma. This chapter, then, looks at the architecture of her works—literally. Sayers's descriptions of rooms and buildings show how her writing performances were often staged according the scripts of her day, revealing a literary aesthetic shaped by modernist sensibilities.

In contrast, the last three chapters of *Writing Performances* demonstrate how Sayers explicitly defied modernism. Chapter 4 makes parallels between Bakhtin and Sayers most explicit, arguing that, even though Bakhtin was a rigorously trained philosopher, celebrated by the theorists of our day, while Sayers was an unabashed popularizer reviled by many theorists of her day, their literary criticism, arising out of an incarnational theology, is very similar.

The next chapter discusses theological issues more fully, demonstrating how Sayers subverts religious language throughout her career in order to construct an unorthodox kind of orthodoxy. Chapter 5 thus argues that Sayers's passionate commitment to Anglo-Catholic Christianity, in combination with her ironizing of faith vocabularies, parallels the paradoxical moves made by many postmodern theorists.

The concluding chapter focuses on a rarely discussed book, *Begin Here*, to demonstrate Sayers's prescience as she called into question the "Absolutes" of modernist culture. Though *Begin Here* was written as an encouragement for the British as they entered World War II, it presents paradigms consonant with those developed by preeminent theorists of postmodernity: Jean-Francois Lyotard, Michel Foucault, Roland Barthes, Jacques Derrida, Judith Butler. However, this chapter will also demonstrate how Sayers was able to avoid the pitfalls often associated with postmodern discourse.

Significantly, the writings of postmodern theorists influenced the development of contemporary performance studies. As Elin Diamond notes,

> Poststructuralist conceptions of the human subject as decentred by language and unconscious desire, and postmodern rejections of foundational discourses (especially totalizing conceptions of gender, race, or national identity) have all made performance and performativity crucial critical tropes. . . .[19]

Performance draws attention to the positionality of perception—to the idea that interpretations of meaning are always situated in a particular theater of representation. In the last half of her career, Sayers explored the theater of Christian representation, often by writing literal scripts that were performed by professional actors. Doing so, she recognized that performance "is precisely the site in which concealed or dissimulated conventions might be investigated."[20]

Some readers may question my alignment of the politically conservative Anglo-Catholic Sayers with postmodern thinkers, considering my climb onto the fountain of critical theory merely a showy opportunity to dive into shallow waters. I take the plunge, therefore, in the same spirit with which Sayers dove into Dante. Just as she believed the fourteenth-century writer to be "much more 'modern' in many respects than the Humanists of the intervening centuries" (*Ltrs* 3. 251),

I make the case that Sayers is much more postmodern in many respects than modern. At the same time, I acknowledge that "modernism" and "postmodern" are problematic terms, encompassing multiple phenomena.[21] As Sayers notes about the "Middle Ages" of Dante, "No age is simple and unitary, but full of cross-currents and bewildering anomalies" (*Ltrs* 3. 247).

Dorothy L. Sayers herself is not simple and unitary. Instead, she stages multiple writing performances in order to address the cross-currents of life. In *Murder Must Advertise* she creates a character in her own image, having Peter Wimsey perform plural roles in order to investigate bewildering anomalies within an advertising agency. When he exits Pym's Publicity after solving the murder, Peter notices that pasted next to a billboard proclaiming "Nutrax For Nerves" is a cigarette slogan he composed while under cover as a copywriter. Sayers, this book will argue, seems to recognize that writing is always undercover—under the cover of discourse, which inevitably disguises truth. Nevertheless, Sayers might well identify with Wimsey as he surveys his writing performance: "He contemplated his work with a kind of amazement. With a few idle words on a sheet of paper he had touched the lives of millions" (355–56).

Chapter 1

The Performance Begins Her(e): The Auto/biographical Sayers

[T]he writing of autobiography is a dangerous business; it is a mark either of great insensitiveness to danger or of an almost supernatural courage.

—Sayers, *The Mind of the Maker*

In *Thrones, Dominations*, a Peter Wimsey novel that Sayers never finished, a husband watches restlessly as his wife awaits the completion of her portrait. The painter complains, "He looks at my picture, and he looks at his wife, and that puzzles him. He looks at her again a long time, and he does not know that what he sees is not there at all." The husband has projected an idealized image onto his wife, an artificial representation that she believes in as well, for "when she looks in the mirror, she too does not see at all what is there."[1] Sayers implies that perceptions can be shaped by expectation, such that seeing itself becomes a scripted performance. She makes the same point in *The Documents in the Case*, where an artist paints two very different women, both of whom do "not see" how they appear to others (53, 56). Here, however, the artist does "not see" as well, failing to recognize how the paintings reveal his own prejudices about the two women.

The problematics of portraiture are analogous to those of biographical and autobiographical endeavors: seeing the other, like perusing the self, is not an innocent activity; our perceptions often perform according to pre-established scripts. Sayers's awareness of this issue helps explain why she didn't "approve much of biographical details" (*Ltrs* 3. 193). She anticipated what her biographers would later demonstrate: that even historical details are subject to writing performances—performances that stage rather than reflect "reality."

Sayers's own attempts to create an autobiographical portrait fueled her suspicions about the authenticity of portraiture, whether rendered on a canvas or in a text. In January of 1932, the year before her fortieth birthday, the author of nine detective novels decided to write her memoirs. After completing thirty-six handwritten pages of "My Edwardian Childhood," Sayers abandoned the project in less than a month, more interested in plans for another Lord Peter Wimsey novel (*Ltrs* 1. 319–20). Two years later, probably after *The Nine Tailors* was finished, she began again with the same autobiographical material; only this time she replaced the first person "I" of "My Edwardian Childhood" with a third-person narrative about Katherine Lammas. Sayers dropped the parallels with her own life as Katherine aged, finally abandoning the novel after 245 handwritten pages. Called *Cat o' Mary*, this manuscript, like the earlier memoir, was never published, perhaps because Sayers recognized that "she too does not see at all what is there."

Theoretical interrogations of autobiography multiplied at the end of the twentieth century, largely due to the poststructuralist assumption that "reality" is an effect of discourse. "Autobiography, then," as Robert Elbaz notes, "is not and cannot be the description of a finished reality, for reality itself is a process of production."[2] Sidonie Smith describes the process of autobiographical production as "storytelling," what we might call a "writing performance":

> [M]emory leaves only a trace of an earlier experience that we adjust into story; experience itself is mediated by the ways we describe and interpret it to others and ourselves; cultural tropes and metaphors which structure autobiographical narrative are themselves fictive; and narrative is driven by its own fictive conventions about beginning, middles, and ends. Even more fundamentally, the language we use to "capture" memory and experience can never "fix" the "real" experience but only approximate it, yielding up it own surplus of meaning or revealing its own artificial closures.[3]

Lest this seem too cynical, even Augustine, whose *Confessions* is often cited as the earliest example of autobiography, admitted that "When we describe the past, it is not the reality of it we are drawing out of our memories, but only words based on impressions of moments that no longer exist."[4]

The problematic relationship between words and memory helps explain Sayers's abandonment of her autobiography soon after she began it. In January of 1932, she tells her cousin Ivy, "I started to write memoirs of my childhood—but I find I don't remember much about it."[5] Instead, she finds that when she writes of childhood she tends to perform her adult self, as seen in her statement, "I dramatised myself, and have at all periods of my life continued to dramatise myself" (EC 11). Sayers seems quite conscious that autobiography, though ostensibly focused upon the past, really serves to explain the present; hence, she repeatedly wonders in her memoir how Freudians might interpret the incidents she describes.

Many of the childhood memories that Sayers puts together, or *re*-members, anticipate an identity defined by writing performances. For example, halfway through "My Edwardian Childhood," Sayers states, "[M]y imagination is purely literary; my fantasies are, and have always been, deliberate literary creation" (12)—and she proceeds to deliver a two-page disquisition on the relationship between fantasy and imagination that will later be transcribed and refined as "Note 'A' " in *The Mind of the Maker.* Doing so, Sayers illustrates the oft-quoted insight of theorist Mikhail Bakhtin:

> I remember my own lived experience in an axiologically active manner not from the aspect of its factually existing content (taken in isolation), but from the aspect of its to-be-attained meaning and object, that is, from the aspect of that which had given meaning to its appearance within me. And in so doing, I renew the still-to-be-achieved character of every one of my experiences, I collect all of my experiences, collect all of myself not in the past, but in the future that confronts me eternally as a future yet-to-be.[6]

Indeed, when Sayers writes "My Edwardian Childhood" she literally inscribes a future yet-to-be, for *The Mind of the Maker* will not be published until nine years later.

Thus, just as Mr. Dick in Charles Dickens' *David Copperfield* cannot keep the head of King Charles I out of his memoirs, so Miss Sayers cannot keep her present head out of the story of her past. Significantly, she gives this very insight to Lord Peter in *The Nine Tailors*, the novel that supplants "My Edwardian Childhood." As Peter puzzles over the authorship of a note found in the belfry of Fenchurch St. Paul, it occurs to him that it might be written by a simpleton named Potty, but since the note does not refer to Potty's favorite subject, hanging, he decides "it is not his—he surely couldn't keep King Charles' Head out of it so long."[7] All writing performances are inevitably colored by opinions, beliefs, and assumptions, often scripted by culture, which artificially construct continuities between past and present. There is hence no unmediated access to the "I" of the past.[8] Bakhtin says of anyone who seeks to write a memoir,

> Even had he created an autobiography or a confession of the most astonishing truthfulness, all the same he, as its creator, remains outside the world he has represented in his work.... It is just as impossible to forge an identity between myself, my own "I," and that "I" that is the subject of my stories as it is to lift myself up by my own hair.[9]

Sayers, who occasionally suffered hair loss when she got under stress, would have appreciated the analogy. She certainly shared Bakhtin's sense that one's past self is always a construction of the present. In a chapter of *The Mind of the Maker* entitled "The Energy Incarnate in Self-Expression," she acknowledges that "autobiography is a book like any other; all the ordinary rules of composition apply to it":

> It is unique, because the author appears, personally and without disguise, as a character in his own story; but it is still a story that he is writing, and he is obliged to handle his own character *as* a character throughout the succession of events. (*Mind* 88)

Sayers, here, provides a clue as to why she replaced the presumably factual "I" of "My Edwardian Childhood" with the apparently fictional character of Katherine in *Cat o' Mary*. As Bakhtin observes,

> [T]he author must take up a position outside himself, must experience himself on a plane that is different from the one on which we actually experience our own life.... He must become another in relation to himself, must look at himself through the eyes of another.[10]

Sayers becomes "another" in relation to herself when she presents experiences of her childhood through the eyes of someone *outside* of Katherine Lammas: a third-person narrator.

Transplanting incidents and passages from "My Edwardian Childhood" into the fictional *Cat o' Mary* two years later, Sayers also changes the structure of her plot, beginning *Cat o' Mary* with dialogue in *media res*: " 'Is it all right?' inquired Mrs Lammas, faintly" (25). More than half of Sayers's Peter Wimsey novels and stories open like this—with dialogue that makes sense only when one continues reading—showing it to be one of her favored approaches to fictive narratives.

In contrast, she begins her earlier memoir with the self-summarizing convention of autobiography: "I am a citizen of no mean city." The fact that she replaced this structure with the dialogue of fiction may indicate that Sayers was aware that the "I" of autobiography is already a fictional writing performance: a projection of the present self onto the vague and inaccurate memories of the past. As Sidonie Smith notes, "the very 'I' of the autobiographer, the apparent sign of a unified 'self' or 'individual,' is always already a . . . fiction of the 'real' and 'true.' "[11]

Changes and additions in *Cat o' Mary* signal Sayers's self-conscious struggle with memories of the "true." In the earlier memoir, Sayers relishes Bluntisham Rectory to which the Sayers family moved in 1898, praising its features with phrases like "personal charm," "agreeable mystery," "beautiful" (20–21). *Cat o' Mary*, however, adds negative comments about the house: it is a "nightmare" echoing with "the sound of falling plaster"; it is a "waste of stone passages and frosty corridors" with no electricity to illuminate the mice scampering within its walls (31). While in the earlier autobiography the kitchen range was "a cavernous affair, given to 'eating coal' " (22) in *Cat o' Mary* it is more malevolently rendered as "a vast, gloomy monster, . . . avid of coal" (31). Significantly, the handwritten page where these bleak descriptions occur is filled with cross-outs, as though Sayers struggled in her presentation of a different vision than the daydream-like setting of "My Edwardian Childhood." She gives as the source of this alternate perspective Mrs. Lammas, a mother aware of the problematic exigencies of life.

The contrast in *Cat o' Mary* between mother and daughter responses to Bluntisham Rectory hints at why Sayers abandoned "My Edwardian Childhood." She recognized how difficult it is to re-present childhood—not only due to the insufficiencies of memory, but also because a child's sensibilities are too far removed from the perspectives of an adult. When a writer tries to re-produce the rarified atmosphere of childhood reveries, she can't help thinking about the suffering yet to come, of adult experiences which destroy childhood dreams.

Recognizing that "childish memory became drowned and lost in later knowledge" (*Cat* 26), Sayers even became tentative about the "facts" themselves. While in "My Edwardian Childhood" she describes herself being horrified when her nurse throws a pneumonia jacket in the fire (9), she is less sure about the agent of the action in *Cat o' Mary*, writing "somebody (Nurse? Mother? the Hospital Nurse?) had lost patience and flung it bodily into the fire" (25–26). Other times she alters the "facts" altogether. Describing her toy monkey Jacko, she says in her memoir that she cut off his chest hair with scissors (EC 13), whereas in *Cat o' Mary* his hair was "worn off his chest by constant cuddling" (28). Which do we believe? The earlier because the text conveys the autobiographical "I"? Or the later, since it has the benefit of reflection and retrieval of memory? Or could the change in *Cat o' Mary* reflect the author's desire to fictively create some endearing qualities for a character who, as the manuscript subtitle tells us, eventually turns into a "prig"?

Sayers may have given us a way to answer such questions within *Cat o' Mary* itself. At one point, Mrs. Lammas asks her daughter how she can write a poem about the dawn when she has never seen it. Katherine admits that she learned about the dawn from books, "but what she had read was to her always just a little more real than what she had experienced." Her perceptions, performing according to the writings she read, generated her own writing performances. Sayers's narrator then comments on Katherine's sense of the "real": "To start with invention is the mark of the fertile mind and leads later to the interpretation of experience; to start with the reproduction of experience is the infallible index of a barren invention."[12] Sayers may have regarded "My Edwardian Childhood" as a barren attempt to *reproduce* experience, whereas *Cat o' Mary* provided her with the fertility of fictional invention that can better *interpret* experience. Perhaps the substitution of *Cat o' Mary* for "My Edwardian Childhood" signals Sayers's recognition that history, even of the self, is always a fictionalized rendering of barren facts, situated upon memories that can only be accessed through the less-than-innocent, though highly fertile, medium of language. As Sayers stated in an essay called "Creative Mind" several years later, "Words . . . are like magnets charged with power that affect and deflect all the instruments of precision that come within their field of influence" (*Whimsical* 93).

It is ironic, then, that Sayers's biographers rely heavily upon texts which Sayers recognized as problematic writing performances. James Brabazon's 1981 biography treats *Cat o' Mary* as a transparent window on Sayers's life and thought, saying in his introduction, "I have made considerable use of the manuscript of *Cat o' Mary*, having convinced myself that at all points it corresponds to Dorothy's own career." David Coomes, who published his biography eleven years after Brabazon, actually says, "Dorothy Leigh Sayers *was* Katherine Lammas." Barbara Reynolds, who knew Sayers quite well, warns us in her 1993 biography that *Cat o' Mary* is not an innocent text, that "it would be a mistake to identify Dorothy and Katherine in every respect."[13] But how does she know in which respects they can be identified? How can any of us assess when an auto/biographical portrait mirrors or merely fabricates experience? After all, as Sayers shows in *Thrones, Dominations*, even a person quite intimate with the subject of a portrait, such as a husband, "does not see at all what is there."

In *The Documents in the Case*, the painter Lathom does not see how much of himself he has put into his portrait of Mrs. Harrison, whom he knew intimately. Munting, who shares a flat with the painter, describes the portrait as "revealing"; we later realize that Lathom's work reveals his erotic relationship with Mrs. Harrison: "only Mrs. H. doesn't see that, and I don't think Lathom realises it either" (*Doc* 53). Significantly, Munting is in the midst of writing a biography as he makes this comment. Struggling with portraiture in another medium, he describes his task as a kind of performance: "My God, but I'm played out! I've been sticking to the accursed *Life* like a leech" (34). Sayers was surely contemplating parallels between the performance of portraiture and that of biography, for, as she was crafting Munting's character in *The Documents in the Case*, she was gathering materials to write a *Life*: the biography of Wilkie Collins. Significantly, in her *Life*, she establishes a parallel between painting and writing: "when Wilkie turns from cataloguing canvasses to describing scenes from nature he can show a pretty skill in word-painting."[14] As with her autobiographies, Sayers never finished her word-painting of Collins; instead, she placed into her novels references to the problematics of portraiture. Sayers seemed to recognize that biographies, like autobiographies, are writing performances influenced by scripts about which Lathom-like authors are unaware.

Biography, like any historical writing, depends not just upon "what is there," but upon the way facts are constructed into meaningful narratives, which inevitably foreground some facts, like details in a painting, as more important than others. However, while a painting captures its subject into one stilled moment, a biographical project depicts the passage of time, ordering facts into cause-and-effect relationships and connecting them to the social contexts in which they are found. Arranging, foregrounding, ordering, and connecting are based upon interpretive paradigms, upon seeing what one expects or wants to see. As Peter Wimsey says of Inspector Sugg in *Whose Body?*, "He didn't find any signs on the roof, because he wasn't looking for them." Sayers correlates this problem with theological issues, having Peter tell Parker that Sugg's investigative tactics are like those of Bible commentators: because they "work with a bias in their minds, . . . they find what they are looking for."[15] Biographers, as well, work with a bias in their minds, often finding what they set out looking for, as can be seen with commentators on Sayers.

Poststructuralists, of course, would deny that we could ever "find" the real Dorothy L. Sayers, for "reality" is always an effect of discourse. This does not mean that the "real" does not exist; it just means that it is always mediated to us through language, which is a human construction. As Linda Hutcheon observes in *The Politics of Postmodernism*,

> To say that the past is only known to us through textual traces is not the same as saying that the past is only textual, as the semiotic idealism of some forms of poststructuralism seems to assert. This ontological reduction is not the point of postmodernism: past events existed empirically, but in epistemological terms we can only know them today through texts. Past events are given meaning, not existence, by their representation in history. (81–82)

Believing that "language is as elusive and disconcerting as any other living thing" (*Ltrs* 4. 115), Sayers would feel quite comfortable with Hutcheon's postmodern approach to history.

In his groundbreaking book *Metahistory: The Historical Imagination in Nineteenth-Century Europe*, Hayden White discusses how meaning is made of history, explaining that historians arrange the facts available to them into

> a hierarchy of significance by assigning events different functions as story elements in such a way as to disclose the formal coherence of a whole set of events considered as a comprehensible process with a discernible beginning, middle, and end. (7)

In other words, *how* historians plot their facts will affect the meaning, the *why*, of historical events.

The relationship between HOW and WHY is addressed several times by Sayers (who usually places the words in upper case) as she reflects on Peter Wimsey's process of detection:

> What nearly always happens to Peter is to collect all kinds of facts, usually in a mood of hopeless bewilderment, until, quite suddenly, the essential facts of the HOW arrange themselves to form a synthesis (by which time, of course, they usually include the WHY as well). (*Ltrs* 1. 389)

Sayers explicitly establishes a parallel between historical scholarship and the process of detection in her most complex novel, *Gaudy Night*, written, it would seem, during (or soon after) her struggle with *Cat o' Mary*. In the novel, Harriet Vane has returned to Oxford to help solve a mystery at her alma mater, Shrewsbury College. Once there, she decides to pursue her own historical project—a study of the nineteenth-century Irish novelist Joseph Le Fanu—thus working simultaneously on the history of a mystery and the mystery of history. As a new researcher at Shrewsbury, Harriet D. Vane parallels the college's new Research Fellow, Helen de Vine—both in activity and in name. Furthermore, Miss De Vine's scholarly investigations are indirectly responsible for Miss D. Vane's detective investigations. The former discovered that an historian named Arthur Robinson had suppressed evidence undermining his M.A. thesis. De Vine's "detective ability," as Peter Wimsey describes it (*Gaudy* 411), caused Robinson to take his life, motivating his distraught widow to harass the school dons with the malevolent and destructive activities that challenge Harriet's "detective ability."

Robinson's crime is the temptation of every historian (and probably most detectives): protection of the paradigm by which one gives meaning to one's discoveries. In a helpful psychoanalytic reading of *Gaudy Night*, Donald G. Marshall describes Robinson with words relevant to any biographer:

> He wishes to be—or at least to seem to be—an historian; and he wishes his theory to be—or at least to seem to be—history. He must conceal his attachment both from the scholarly community and from himself, must perversely repress his real feeling for his creation beneath an apparent detachment.[16]

Today, of course, the Enlightenment mystification of scholarly detachment has been exposed, like Robinson himself, to be inaccurate. Complete detachment is impossible; all histories are writing performances following scripts that explain, rather than merely present, the facts. This, of course, does not justify the destruction of evidence; empirically verifiable facts, even for the postmodernist, as Linda Hutcheon has made clear, must still be acknowledged.

Wanting to protect the validity of facts, Hayden White distinguishes between "chronicle" and "story" in the plotting of history. A "chronicle" merely places events in chronological order, with no allusion to their significance, as can be done with Sayers's life:

> A Chronicle of Sayers's Life
>
> Born in Oxford on June 13, 1893, Dorothy Leigh Sayers, an only child, lived with her parents at Bluntisham Rectory, Cambridgeshire, before attending boarding school in Salisbury, where she won a scholarship to Somerville College, Oxford. After graduating with first class

> honors in French, she briefly worked as a teacher, an apprentice at Blackwell's publishing firm, which produced two books of her poetry, and an administrative assistant for a school in Normandy. From 1922 to 1931 she was employed as a copywriter at Bensons, a London advertising firm. During that time, she gave birth to John Anthony in 1924 and married Arthur Fleming in 1926. Sayers published twelve detective novels between 1923 and 1937, the year in which she wrote a play for the Canterbury Festival. She wrote five more plays on Christian topics, two for B.B.C. radio, which led to numerous talks and radio spots about theological and literary issues, including a book-length study in 1941 about the relevance of theology to literary criticism. In the last decade of her life, Sayers focused most of her attention on translating and annotating Dante's *Divine Comedy* for Penguin. She died of a coronary thrombosis on Dec. 17, 1957.

This bare-bones factual portrait, however, would look different if painted by anyone else, who surely would choose alternate facts to include, like Sayers's translations of Medieval French narratives, her two secular dramatic productions, her friendships with Charles Williams and C. S. Lewis, or her involvements with The Detection Club and St. Anne's House, a center engaging Christianity and contemporary culture. The biographer's vision affects her version.[17]

If even the selection of facts reflects a biographer's biases, how much more so might the plotting of them into narrative. According to Hayden White, the process of historical construction necessitates "spectacle," what I would call a "writing performance": "the chronicle is organized into a story by the further arrangement of the events into the components of a 'spectacle' or process of happening" (5). In *Gaudy Night*, Sayers has Peter Wimsey discuss the importance of storytelling to the (re)construction of history. After reading Harriet's account of the malicious activities at Shrewsbury, he compliments her with, "I'll say one thing for the writing of detective fiction: you know how to put your story together; how to arrange the evidence" (321). Harriet's ability as a detective fiction writer enables her to write an effective history of factual events. Significantly, Hayden White regards the detective genre as one of the plot structures historians employ to give meaning to facts: "Precisely because the historian is not (or claims not to be) telling the story 'for its own sake,' he is inclined to emplot his stories in the most conventional forms—as fairy tale or detective story on the one hand, as Romance, Comedy, Tragedy, or Satire on the other" (8, nt. 6).

Sayers's biographers plot their subject's life according to similar narrative strategies. Janet Hitchman, whose *Such a Strange Lady*

(1975) revealed to the world that Sayers had secretly given birth to a son, embellishes her unfootnoted biography with imaginative dialogue and drama, garnering disdain from later biographers. James Brabazon, who wrote the first biography to be authorized by Sayers's son (1981), uses the language of portraiture to describe the inaccuracies of Hitchman's text: "The picture that has been drawn of Dorothy besottedly creeping around Oxford in the footsteps of [her friend Eric] Whelpton, making his life a misery with unexpected and unwelcome overtures, appears to have no foundation whatever" (71). Several years before Brabazon, Ralph Hone redressed Sayers in a 1979 literary biography, foregrounding the productions of Sayers's mind over the performances of her body. However, even he could not help speculating about the father of Sayers's child, using a boldly inaccurate brushstroke: "Undoubtedly Dorothy believed [John] Cournos was the father of the child."[18] A year later, Nancy M. Tischler redeems Sayers using yet another kind of plot: the allegory of *Pilgrim's Progress*. Naming her biography *Dorothy L. Sayers: A Pilgrim Soul*, Tischler employs chapter titles like "Vanity Fair" and "The View from the Delectable Mountains" to present the various stages of Sayers's life. "Like Bunyan's pilgrim," Tischler writes, "Dorothy L. Sayers went through the Valley of Humiliation and confronted Apollyon" (55) until she triumphantly crossed over to "The Celestial City," as Tischler titles her final chapter. She scripts Sayers's life in the Romance genre, described by Hayden White as

> the hero's transcendence of the world of experience, his victory over it, and his final liberation from it.... It is a drama of the triumph of good over evil,...and of the ultimate transcendence of man over the world in which he was imprisoned by the Fall. (9)

White notes that the transformation of chronicle into historical plot necessitates the formulation of a beginning, middle, and end to the story—whether one of Romance or detection. How a biography starts, then, often signals the way the author views her subject. Significantly, Janet Hitchman begins with a two-page fiction: "Catherine," an Oxford University tour guide, has trouble generating any response from two visitors from South Africa until she shows them the bridge "where Lord Peter Wimsey and Harriet Vane became engaged."[19] As the couple suddenly gets excited about Oxford, Hitchman comments that sometimes fiction is more "real" than fact. In contrast, perhaps intentionally, Ralph Hone and Barbara Reynolds begin their biographies with the hard, cold fact of Sayers's birth, both emphasizing the growth of her genius as they plot her growth from child to adult. David Coomes takes an entirely different tack, starting his biography with an account of letters exchanged in 1940 between Sayers and the B.B.C. In his preface, Coomes makes his purposes quite clear: "Other biographers, notably James Brabazon, have concluded that [Sayers's] faith owed more to the head than the heart, that she merely exulted in the stimulus of an intellectual godgame. I believe her letters prove otherwise" (8). Hence, he begins with his proclaimed source of evidence: letters expressing what he takes to be Sayers's heartfelt response to writing about Jesus in *The Man Born to Be King*.

Brabazon begins with correspondence as well, but to a very different end. He quotes a letter in which the adult Sayers complains about not being able to get what she wants, including a "man." Within his first dozen lines after the quotation, Brabazon uses such negative terms as "lack," "scarce," "impossible," "obsessed," "lowest point of... disheartenment," "worse things," and "long-drawn-out drag," employing the word "sex" twice as he tells us that Sayers was a twenty-eight-year-old virgin (1). Later, he uses the gender-inflected word "hysterical" (from the Greek word for "womb") to describe her reaction to proposed alterations in her script for *The Man Born to Be King*: "[I]t was the staff of the BBC who behaved like professionals, and Dorothy who behaved like a spoilt and hysterical amateur, reinforcing her tantrums by an extremely dubious application of religious doctrine" (200).[20] An interrelationship between frustration, obsession, and sexuality is embedded in Brabazon's diction, perhaps because he plots Sayers's life as a tragedy. According to Hayden White, "In Tragedy,... there are intimations of states of division among men more terrible than that which incited the tragic agon at the beginning of the drama" (9).

For Brabazon, Sayers's tragic agon begins with the birth of her son, John Anthony. In his closing chapter he states, "I cannot help regretting, however uselessly, that the events of the early twenties worked out as they did for her; and that at a crucial stage in her life and career she was saddled with so much misery and with a life-long secret that cannot have failed to affect her" (275). Brabazon then indicates that Sayers might have produced far greater work if it weren't for her secret agon; he doubts whether her Festival plays "will survive," he thinks her translations "will last only a limited time," and he regards most of her essays as "concerned with matters of passing moment" (276). It is as though Sayers's tragic creation of an illegitimate child will last longer than most of her other work.

In her 1993 centenary biography, Barbara Reynolds plots Sayers's life in contradistinction to Brabazon, explicitly arguing against or qualifying his interpretation at least seven times in her text. In contrast to his tragic emplotment, her history reads like Comedy, which White describes as "the temporary triumph of man over his world by the prospect of occasional reconciliations of the forces at play in the social and natural worlds" (9). Indeed, at one point in her text Reynolds writes, "Like Harriet Vane, Dorothy possessed the exuberance to rise above her constraints and difficulties" (231). While Brabazon berates Sayers for her hysterical tantrums over her script for *The Man Born to Be King*, Reynolds dismisses the incident: "The misunderstanding with the B.B.C., though distressful at the time, was no more than a minor skirmish, soon in any case to be resolved" (307).

White explains that the "reconciliations" imaged by Comic emplotments are symbolized by "festive occasions," whereas in Tragedy "there are no festive occasions, except false or illusory ones" (9). True to her emplotment, then, Reynolds writes that Sayers "enjoyed gala occasions, such as dinners of the Detection Club, for which she dressed with splendour" (361). Brabazon, in contrast, dourly states of a Detection Club ceremony, "she treated the whole thing with such solemnity as to deprive it of much of its fun," summarizing several lines later, "life persisted in reminding her of her fallibility, by working out contrary to her hopes" (145). While Brabazon closes his biography emphasizing Sayers's fatal fallibility, Reynolds finishes as though she had read White's reference to the "festive occasions" of Comedy: "[Sayers] was linked to her friends by a vast network of letters, by many festive occasions, and, above all, by work shared and undertaken in common" (371). Significantly, Reynolds ends her plot of Sayers's life using a word that we all associate with Comedy: "laughing."

The differing emplotments of Reynolds and Brabazon become quite apparent as they describe Sayers's infatuation with and despair over John Cournos, a journalist, poet, and novelist whom she met in 1921. Unlike previous biographers, Reynolds and Brabazon had access to personal documents others did not see: the letters of jealous frustration Sayers wrote in 1924 after she heard that John, the avowed bachelor, had married (*Ltrs* 1. 215–39). In these painful letters, Sayers openly confesses to Cournos not only her misery over the way he treated her, but also her illegitimate baby—a confession heard by no other human at that time except, perhaps, for Ivy Shrimpton, the cousin who raised John Anthony.

The biographer's access to private letters parallels a quandary that Sayers created for a character in her third detective novel. In

Unnatural Death, Miss Climpson becomes privy to a document not meant to be known by any but the sender and the receiver: notes for a confession. When Miss Climpson realizes the paper dropped outside a church confessional was written by Vera Findlater, a woman she has befriended while investigating a case for Lord Peter, Miss Climpson agonizes over whether to read it. Finally, with "a red flush creeping over her cheeks as though she were stripping something naked," Miss Climpson glances at the words "Jealousy" and "Idol," clues to Vera's abject worship of another woman (*Death* 551). Sayers, in her own revelatory letters written a couple of years before *Unnatural Death*, describes herself feeling similarly about Cournos: "I suppose what I had for you was one of those sort of abject hero-worships" (Ltrs 1. 223). Furthermore, Miss Climpson's response to the confessional note is very similar to that of Sayers's closest friends when they learned about John Anthony only after Sayers's death: " 'Queer,' thought Miss Climpson, 'that Vera has never come and told me about her trouble. But perhaps she is ashamed, poor child' " (*Death* 552).

Brabazon and Reynolds both knew Sayers, the former when they briefly served together as churchwardens in the same London parishes, and the latter in a much more significant relationship developed through a common love of Dante. Both, like Miss Climpson of Vera Findlater, sought to interpret Sayers's letters of confession. Not surprisingly, Brabazon detected Sayers's tragic flaw, her "human weakness exposed" through the letters:

> It is not simply that they show that Dorothy shared the frailties of the rest of humanity; their passion and their pain also go a long way towards explaining the rigidity with which, once the affair was over, she protected her private self against further risk, and retreated into an invulnerability in which attack was her best means of defence. (89)

Brabazon emphasizes the negative consequences of Sayers's frailty: her rigidity and invulnerability, her proclivity to attack. Reynolds, however, chooses to discount Sayers's embarrassing confession to Cournos. Though she believes Sayers usually "writes of herself unfeignedly," Reynolds states that in these letters Sayers "was playing a part of being unhappy, or trying to find the role which would best help her to regain her balance." Though this statement appears in her introduction to the first volume of Sayers's letters (*Ltrs* 1. xviii), the attitude behind it certainly informed her Sayers biography published two years previously. To maintain her Comedy, Reynolds must interpret Sayers's despairing letters as merely an act: a writing performance.

Brabazon also interprets Sayers's correspondence as an act—when it suits his sense of Tragedy. While Reynolds believes that Sayers "was playing a part of being *un*happy" with Cournos, Brabazon believes that her adolescent self was playing a part of being *happy* while at the Godolphin, her boarding school in Salisbury. He bases his theory upon a letter Sayers wrote twenty years after boarding school, in which she discusses whether the six-year-old John Anthony should be baptized. Referring to her own confirmation while at Godolphin, Sayers advises her cousin Ivy, "Being baptised without one's will is certainly not so harmful as being confirmed against one's will, which is what happened to me, and gave me a resentment against religion in general which lasted a long time" (*Ltrs* 1. 306). Barbara Reynolds, not surprisingly, states there is no evidence from Sayers's adolescent correspondence that she resisted her confirmation while at the Godolphin school or subsequently held "a resentment against religion" (*Ltrs* 1. 307, nt. 2). Indeed, after her confirmation, Dorothy not only wrote her parents about the "awfully nice" and "beautifully arranged" service (*Ltrs* 1. 40), but also told Ivy "[i]t was a lovely day, and it all went off beautifully" (41). Brabazon dismisses these statements as signs of a false writing performance, referring instead to Katherine's misery over being confirmed in *Cat o' Mary*. Because it fits his tragic emplotment, the fictional *Cat o' Mary* performs Sayers's thoughts better than the facts she writes in letters immediately after her confirmation. Thus, while taking the disconsolate letters to Cournos at face value, Brabazon interprets Sayers's cheerful Goldophin letters as a disguise for "mental and physical distress": a pretense of happiness for the sake of her family's feelings (33).

Brabazon's belief that the young Dorothy wanted to protect her family from distress, though consonant with statements in *Cat o' Mary*,[21] fails to account for explicitly negative references in Sayers's Godolphin letters, as when the schoolgirl complains of a "loathly" teacher, a "deadly dull" term, or her distaste for the "peculiarly ugly sort of chant" by which the Athanasian Creed is delivered in church (Ltrs 1. 36, 37, 43). Brabazon, it would seem, chose to believe Sayers's 1930 letter to Ivy rather than her earlier schoolgirl effusions because doing so better fufills his tragic script. In fact, he begins his narration of the confirmation service with "The worst of it was that at the moment of her greatest need, her parents too abandoned her" (35). This is somewhat excessive language to describe the fact that Sayers's parents wanted their daughter to be confirmed (a point which Reynolds actually contests [*Life* 39]).

A similar disparity in emplotments occurs between biographical accounts of Sayers's rebound relationship with Bill White, the father of John Anthony.[22] Both Brabazon and Reynolds quote from a letter Sayers sends her parents about an evening when she and Bill slogged through fog and mud to join some of her friends for a night of dancing. The friends, however, didn't want to venture out in such bad weather, causing Sayers to tell "the whole lot of them that they were damned unsporting" and to contemptuously refer to them as "sleek and oily" (Ltrs 1. 200). Brabazon interprets this incident, of course, as evidence of Sayers's tragic flaw—her need for a man—causing her to compromise her values:

> Never before or after did Dorothy evince any preference for cold fresh air over warmth and fug, . . . any dislike of sleekness or upper-class voices. . . . The whole passage reeks of the contempt of the man who works with his hands for the man who does not, the softie, the sissy, the talker. It is on Bill's behalf that Dorothy is indignant, it is through his eyes that she sees her erstwhile friends, even some of the words are surely his. (98–99)

While Brabazon reads Bill as Iago-like, getting the Othello-like Sayers to view reality according to his perspective, Reynolds regards Bill as a Sancho Panzo sidekick who teaches the Quixotic hero new ways of seeing: "By now Bill has made her other friends seem effete—'soft-gutted', she called them, perhaps borrowing the phrase from Bill. In contrast, he was virile, tough and foul-mouthed" (120). Reynolds even mitigates his foul-mouth by assuring us that Sayers "was to give this coarseness to her aristocratic Lord Peter," who recites an obscene limerick in *Clouds of Witness* (*Life* 120). This is one of several ways Reynolds elevates Bill in her biography, as opposed to Brabazon, who shows Bill pulling Sayers down to working-class behavior. While Brabazon states, "Dorothy had lowered her sights by now," and "Bill was for Dorothy a second best" (97), Reynolds speculates, with no evidence from the letters, that during a visit to Sayers's parents, "Whatever clothes [Bill] had brought in his saddle-bags, a formal suit, at least, would have been included, if not a dinner-jacket" (118), and she imagines how he may have dressed at a dance sponsored by Sayers's employer, Benson's: "This occasion surely called for a dinner-jacket; possibly, since 'he had not a red cent,' they managed to hire or borrow one" (119).

Just as Reynolds imaginatively clothes Bill's body in a gentlemanly fashion, she does the same with his tongue, quoting a letter Sayers sent Ivy after the birth of the baby: "John's father is, exactly as I hoped, beginning to feel less sore and take a little pride and interest . . . I showed him your last letter, and he said, 'What an awfully nice woman she must be!' " Reynolds then adds her own commentary on Bill:

> Since these are the only recorded words of John Anthony's father, it is perhaps worthwhile to pause and consider what they reveal about him. First, he was decently grateful in his feelings towards Ivy. Secondly, his comment, brief as it is, has the unmistakable stamp of an educated person. In other words, to use the phraseology of the period, he spoke like a gentleman. (144)

Though there's a strong possibility that Sayers may have reworded Bill's statement to make it more pleasing to Ivy, whose conservatism Sayers mentions several times in her letters, Reynolds wants to take Sayers at her word—and at Bill's—to maintain the reconciliatory gestures of Comedy. This wouldn't stand out if Reynolds, at other times, did not question Sayers's account of things, as when she quotes from Sayers's essay "Gaudy Night" (1937): "Sayers said that when she embarked on *Strong Poison* 'it was with the infanticidal intention of doing away with Peter, that is, of marrying him off and getting rid of him' " (229). After giving several arguments for the inaccuracy of this statement, Reynolds summarizes with, "It is doubtful that she was really tired of Wimsey" (230). Like a painter, Reynolds chooses which characteristics she wants to foreground in her portrait.

I am not questioning the scholarly integrity of Reynolds, whom I admire greatly. I am merely illustrating Hayden White's argument, and Dorothy L. Sayers's intuition, that the very act of narrating history becomes a writing performance that creates meaning. Even "facts" are not innocent, as Reynolds shows when she valorizes the "fact" of one statement from Sayers, and discounts the "fact" of another.[23] Both Brabazon and Reynolds choose and shape events in order to fit their modes of emplotment, such that their writing performances follow radically different scripts.

Sayers's biographers are like the Gotobed sisters in *Unnatural Death*, called in to Miss Dawson's sick room by Mary Whitaker in order to sign papers about what they see. However, a screen has been set up between the sisters and the object of their observation, Miss Dawson, forcing them to look at her actions in a mirror. Sayers actually provides a chart in the text to make sure we understand the obliqueness of their vision, labeling the sisters as "witnesses" with "writing materials." Just as the oddly named Gotobed sisters cannot really go to the bed of Miss Dawson, so biographical witnesses, using their own "writing materials," cannot directly access the life of their subjects, for the screens of autobiographical memory, personal prejudice, and discursive practice always separate them from their focus. Or, as Sidone Smith puts it, " 'Experience' is not out there to be recovered outside the interpretive grids of culture and the structural grids of language."[24] Brabazon and Coomes, then, look at Miss Sayers via the mirror of *Cat o' Mary*, signing their names to biographical texts with the assumption that what they see in the mirror is "true." At least Reynolds acknowledges that actions documented in *Cat o' Mary* are often fabrications, just like the fabricated documents that Mary Whitaker, behind the screen, tries to get Miss Dawson to sign.

Sayers herself could not sign off on her own autobiography, realizing that she had no direct access to the person she once was, that memory only obliquely mirrors the past. It is no wonder, then, that she repeatedly disparages biography in her correspondence and eschews personal questions during interviews. In one letter, she gets quite earthy in her deprecation of readers with a "craze for the 'personal angle' ": "They eat men and events and turn them all into wind. And there is nothing that you or I or anybody can do about it, being merely the fodder for their intestinal noises." Rather than the ephemeral fumes of biography, Sayers wanted *her work* to memorialize her: "biographical detail . . . distracts attention from the work to the worker—a bad thing always." Exasperated over having to dodge " 'Profiles,' 'Portraits' and other devices by which the fluttering attention of the public is diverted from what one does to what one looks like," she believed that an author's writing performances are always more important than writing performances *about* an author.[25]

When the artist in *Thrones, Dominations* paints his portrait of Harwell's wife, Rosamund, he diverts fluttering attention from what she looks like by placing something in her hands: a papier-mache mask of her own face that has been created by another painter. While Sayers's fragment ends with this incident, Jill Paton Walsh finishes the novel by having a murderer cover up his crime by placing the papier-mache mask on the face of Rosamund's dead body: another apt metaphor for the performance of biographical writing. One could say that all biographies are lifelike masks that overlay the dead bodies of their subject-matter—some more lifelike than others, though masks all the same.

In his famous article "Autobiography as De-facement," Paul de Man regards the life-writing genre as an enactment of the trope of personification, the formal term for which is "prosopopeia," which literally means "conferring a mask or a face." As Laura Marcus explains it, "the giving of face is also a de-facement; the textual subject is stabilized and given identity through its assumption of face, but to the extent that this is a false front, a face for the faceless, it de-faces."[26] Indeed, when the murderer in *Thrones, Dominations* covers up the disfigured face of Mrs. Harwell, it is a type of de-facement. Even though the mask (re)presents her beauty such that a viewer through the window thinks she is still alive, it does not manifest the truth. As Marcus notes of auto/biography, "In order for the face, in its facelessness, to be recognized, it must be undone, de-faced."[27] When the mask is removed, Rosamund's face is in fact defaced, covered with ugly bruises; but this defacement is more true to her lived experience than is the mask of her beauty.

Some readers may consider my unmasking of Sayers biographies to be a defacement of significant scholarship. My intent, however, is to suggest that the genre of biography, like that of autobiography, inevitably produces a de-facement. Nevertheless, just as painted portraits give us pleasure through their differing interpretations, so also do biographies. Mikhail Bakhtin, in fact, would celebrate their differences. Rather than accede to the modernist paradigm of a univocal "authentic" self that must be discovered, as though it were a mystery to be solved by the most proficient biographical detective, Bakhtin regarded identity as polyphonic. So also, I would suggest, did Sayers.

In *Cat o' Mary*, the very writing of which implies her difficulty with an authentic portrait of the self, Sayers contemplates the performance of identity: " 'Just be yourself', as Mrs. Lammas said. But what was 'self'? You could take off layer after layer, like peeling an onion, and there was always another self down below, watching and mocking." This perspective is contrasted with a conventional, and tacitly problematic, script about identity: "You had your 'real self' (a smooth, round self-contained entity as hard and clear as a crystal ball) and anything superimposed on that was insincerity or showing-off" (127–28). In contrast, Sayers seems to delight in the fact that Katherine often "lost her identity in that of some beloved character in a book." To lose "identity" in such a way might perhaps be a means of finding it: "In bed at night, she lulled herself to sleep with long stories, recited aloud, in which she *identified* herself with some hero of fiction," whom she transported to far away countries and far off times (37, emphasis mine). The self becomes molded by the imaginative recitation of writing performances.

Bakhtin, in fact, established a parallel between writing performances and the self. Just as the truly artistic novel is inscribed with multiple perspectives that cannot be reduced to one dominant ideology, causing different readers to interpret the text in different ways, so the self is marked by "heteroglossia" or "divergent tongues." Some interpreters of the self will understand certain of those tongues while remaining deaf to those that speak to others. Nevertheless, all people outside the self can exercise what Bakhtin calls an "excess of seeing," perceiving things to which the self remains blind. Therefore the biographer, in Bakhtin's perception,

> will seek to consummate the hero's life from a point of view which is in principle different from the way that life was lived and experienced by the hero himself from within himself. Every line written, every step taken by the narrator, will strive, in that case, to utilize the narrator's *fundamental and essential excess of seeing.*[28]

Brabazon sees Sayers as a tragic figure, while Reynolds sees her as a hero of Comedy; Tischler sees Sayers in a Romance while Hitchman sees Sayers's strangeness. All, of course, are right, consummating the various tongues of Sayers's heteroglossic self on different stages of representation. The life of Sayers makes clear that all biographies are writing performances, following significantly different scripts in order to narrate the facts.

Sayers, however, would rather have us study her own writing performances. As she states in her chapter on autobiography, the "mind of the maker is generally revealed, and in a manner incarnate, in all its creation" (*Mind* 87). Sayers saw herself incarnate not in the activities of her flesh, but in the productions of her mind.

Significantly, just a few hours before she died, Sayers went to an exhibition at the Royal Institute Galleries, Piccadilly, in order to peruse a recently hung full-body portrait of herself. One wonders what she saw.

Chapter 2

Identifying Gender(ed) Performances

I was not so sure
I wanted to "identify myself,"
as the phrase goes, with feminism.

—Sayers, "Are Women Human?"

In order to "identify" Dorothy L. Sayers, biographers often parallel her with Harriet Vane, the love interest she created for Peter in her sixth novel, *Strong Poison*, and whose character she developed most fully in *Gaudy Night*. After all, Vane, like Sayers, is a graduate of an Oxford college who writes detective novels, profoundly committed to the integrity of her work. Both have fathers named Henry, and both were deeply hurt by sexual relationships which left them responsible for the bodies of others: Vane for the lover she was accused of murdering, Sayers for the secret child she supported for the rest of her life. James Brabazon calls Vane "the figure based on Dorothy herself," who "bestow[ed] her own attributes on Harriet." David Coomes says Harriet, "a mirror image of Sayers," is "a thinly-disguised autobiographical figure." Barbara Reynolds states that "Harriet is a more credible character than Peter, being in many respects a projection of the author" and she outlines in some detail how the heroine of *Gaudy Night* parallels her creator. However, Reynolds also lists at least fourteen Peter Wimsey characteristics that she recognized in her friend, giving helpful examples throughout her biography of elements from Sayers's life that make their way into her detective fiction. Of twenty-two intriguing parallels Reynolds provides, seventeen are from the pre-Harriet Vane novels, leading us to question why Harriet Vane mirrors Sayers any more than does Peter Wimsey.[1]

The answer may be one of gender: readers seem more comfortable aligning Sayers with a female character than with a male. In their biographies, Hitchman, Brabazon, and Coomes explicitly defend Sayers's heterosexuality, assuring us that, despite her loud voice and "mannish" dress, the creator of Peter Wimsey was not a lesbian.[2] Reynolds tells us that Mac, Sayers's husband, fulfilled her sexually, and she includes an alleged incident from Sayers's relationship with John Cournos: "In their love-making they went so far as to undress and lie naked on a couch together, but no further" (160, 112). Reynolds indirectly provides us with her source for this information: a novel that Cournos published in 1932 entitled *The Devil is an English Gentleman*, in which characters named "Richard" and "Stella" lie naked together on a couch. In the novel, Stella speaks lines that Cournos copied straight out of Sayers's letters to him (!!), leading Reynolds to assume that the incident on the couch happened between Dorothy and John, even though Stella is quite different from Dorothy.[3] Reynolds then suggests that John, and *not* Dorothy, kept the affair from being consummated: "Either he was impotent or he took pleasure in the power that he, the son of impoverished Jewish immigrants, could thus exercise over an Oxford-educated, still virginal daughter of the Church of England" (112–13). It would seem that Reynolds includes these speculations in order to address concerns over the instability of Sayers's gender identity.

Until the last part of the twentieth century, gender and sexuality were regarded as identical, evident in an article published the same year that Sayers was fleshing out Miss Milsom's obsession with sexuality in *The Documents in the Case*. In 1929, Joan Riviere, a psychoanalyst, cited Ernest Jones's definition of certain kinds of "homosexual" women who, "while taking no interest in other women, wish for 'recognition' of their masculinity from men and claim to be the equals of men, or in other words, to be men themselves."[4] These words could easily describe Sayers, who gave no indication of having sexual "interest in other women," but who believed herself to be the equal of men, displaying "a manner," according to C. S. Lewis, "superficially masculine and even gleefully ogreish."[5] Today, we would recognize such egalitarian desires and "masculine" performances not as undermining sexuality, but as qualifying constructions of gender by which a society assumes females to be subservient to males. Gender is performed according to cultural scripts.

Sayers, however, liked performing scripts written for men. With the exception of the excessively feminized pose in a photograph attributed to her Godolphin days (figure 2.1), evidence points to the fact that she invariably played male roles in dramatic productions, as with the

Figure 2.1 Possibly performing in *Coriolanus* at the Godolphin school (Courtesy of The Marion E. Wade Center).

swashbuckling character of Athos from Alexandre Dumas' *The Three Musketeers* (figure 2.2). At age fifteen, she helped set up a performance of "Red Riding-Hood" at the Godolphin school, playing a prince disguised as a wolf, moustache and all. When she put up one of her own plays at Godolphin, she designed the role of "bloody villain" for herself, asking her parents to send costume paraphernalia that would

Figure 2.2 Performing Athos at Bluntisham Rectory (Courtesy of The Marion E. Wade Center).

mark her as a male—boots, brigand hat, swords, and dagger—just as she would later write for a beard, breeches, and false nose in order to play Shylock in Shakespeare's *The Merchant of Venice.* For a ball at Godolphin she dressed like Henry V, and for a dance at Oxford she went as Tristram. While at Somerville College, she took the leading role of John Gaunt in *The Admiral Guinea*, and in the Going-Down Play of 1915 she impersonated Hugh Percy Allen, the conductor of the Oxford Bach Choir (figure 2.3).[6] While these male performances

Figure 2.3 Performing Hugh Percy Allen at Somerville College (Courtesy of The Marion E. Wade Center).

can be partially explained by the fact that Sayers was an only child who attended all-female schools, it is still significant that she almost always seemed to take on a leading male role, at one point writing to her parents, "It surprises everyone that I make up so well as a man. It surprises me rather. . . . " (*Ltrs* 1. 81). This, perhaps, was her first glimmer that gender is performed according to the scripts of culture; indeed, the photographs of her playing Athos and H. P. Allen show her not only in costume, but also acting out hyperbolic poses and gestures that "identify" the character as "male."[7]

Sayers so enjoyed these male performances that she often extended them in her writing performances, signing letters with the names of characters she had played—Shylock, J. G. (John Gaunt), H. P. Rallentando (H. P. Allen)—as well as with "Alexander Mitchingham," a persona she took on to spoof literary criticism, and "Matthew Wimsey," a character she wrote into *Busman's Honeymoon*. She used the name "Roger Wimsey" to sign a series of love poems, ostensibly written in 1585, that she sent to Muriel St. Clare Byrne, her collaborator on *Busman's Honeymoon*. Even as an adolescent, Sayers often performed her correspondence as a male. For nearly six years (from fall 1907 to summer 1913), Dorothy signed many of her letters to Ivy with "Athos," often addressing her cousin as the "Duchesse de Chevreuse" in order to histrionically proclaim her desire: "Oh, love, love! To see thee again! To hold thee once more to my heart! I laid thy letter there—but I would embrace thine own dear self. Oh, my adored one, come! come!"[8] In this unpublished letter, Dorothy self-consciously performs the role of male lover, signaled by the excesses of her prose with its archaic pronouns and multiple exclamation marks. In a similar letter, Dorothy ironizes her sensationally romantic performance with parenthetical asides: "I have made some new songs to sing to you, and we will have all the old ones as well (provided I do not catch a cold)" (*Ltrs* 1. 12). Not only do we perceive here the seeds of Peter Wimsey's histrionic verbal performances, but we also get an insight to the roles the cousins played throughout their lifetimes. Even though Dorothy was eight years younger than Ivy, she performed the conventionally dominant "male" role to Ivy's "female" persona. In a letter written eight to nine months before the lover's lament, Dorothy, in her "own" voice, challenged Ivy to be less judgmental: "you have rather a high moral standard, and if people don't keep up to it, you are a bit apt to condemn them." It is in this letter that Sayers made her uncannily prophetic statement, noted by many of her biographers: "I shouldn't like to feel, Ivy, that supposing some time I sinned a great sin, that I should be afraid to come to you for help" (*Ltrs* 1. 10).

Sixteen years later, Sayers did ask Ivy to take care of her illegitimate baby, perpetuating their gendered performances: Sayers as financial provider, Ivy as childrearer. In a letter written in 1931, when John Anthony was eight years old, Sayers sounds like a typical husband of the day, busy with his professional activities while his spouse is at home with the children:

> Don't be worried about expenses—just let me know what your visits etc. cost you and I will send a cheque along.... I wish I could come along and see you, but I am bunged up with engagements in Town and various bothers connected with books and publishers which need me and which *Mean Money*! (*Ltrs* 1. 314)

Several years earlier Sayers had explicitly aligned her response to children with male sensibilities, telling Ivy, "I don't think small children appeal much to men (nor, indeed, to me)" (*Ltrs* 1. 285).[9]

To interpret from these examples that Sayers might be a lesbian misses the point. For many contemporary theorists, sexuality is ground in biological imperatives, while gender is situated in the performance of cultural imperatives.[10] "Thus," as Mary Eagleton notes, "what our culture understands by feminine behaviour is not the consequence of the product of a feminine identity; instead, our understanding of a feminine identity is produced, within signification, through the repeated performance of words and actions which we code as 'feminine.'" Laurence Senelick tersely summarizes current theories with the statement "Gender *is* performance."[11]

A clear division between the genders, then, is perpetuated by arbitrarily constructed cultural markers: men are logical, women are emotional; the love of motorcycles and cigars (both of which Sayers relished) is male, the love of babies is female. Even though such scripts reinforce that a woman's place is as a nurturer in the home while the male empowers and controls society, both genders tend to affirm their society's constructions as they seek to assume a clear identity. Problematically, such gender performances, written by culture, perpetuate the construction of male superiority, as when it is considered cute for a girl to be a "tomboy," but embarrassing in a boy to be an effeminate "sissy." After all, the girl is mimicking the superior gender, whereas the boy is lowering himself to perform female characteristics.[12]

Sayers challenged cultural constructions of gender in her essays "Are Women Human?" (1938) and "The Human-Not-Quite-Human" (1941), influenced, perhaps, by her education at Somerville College, one of the "hot-beds of feminism" the year she matriculated.[13] In the earlier essay, Sayers asserts that males and females, as equally human, have a right to choose identifying characteristics and roles: "a woman is just as much an ordinary human being as a man, with the same individual preferences, and with just as much right to the tastes and preferences of an individual" (1938, 19). She therefore attacks those cultural institutions that privilege males: schooling, vocation, and fashion.

Sayers, of course, speaks from experience. Even though she took the same courses and passed the same exams as her male counterparts at Oxford, earning a First in French while attending Somerville College (1912–15), Sayers was not awarded a degree. In 1915, it was still considered inappropriate, if not impossible, for women to achieve the same educational expertise as men. In 1920, Oxford changed its policy, granting, for the first time in its history, B.A. and M.A. degrees to five women retroactively, including Sayers. While lauding this change in "Are Women Human?" Sayers warns against women's colleges merely copying male educational paradigms, an action which implicitly reinscribes male superiority. Furthermore, as Sayers notes, male institutions are not worth imitating, for they are "autocratic, old-fashioned and in many respects inefficient" (1938, 21).

Along with the freedom to choose an education, Sayers argues, should come freedom to pursue a vocation in defiance of gendered expectations: "If any individual woman is able to make a first-class lawyer, doctor, architect or engineer, then she must be allowed to try her hand at it" (1938, 25). Sayers had made this point as early as 1926, creating a character named Miss Climpson as an exemplar of female capabilities suppressed by the gendered scripts of culture. When Detective-Inspector Parker meets her, he immediately recognizes Miss Climpson's intelligence, complimenting her "clear head and retentive memory." Miss Climpson's response reveals that society had overlooked her acumen, treating her as human-not-quite-human:

> A dear old friend of mine used to say that I should have made a very good lawyer... but of course, when I was young, girls didn't have the education or the *opportunities* they get nowadays. (*Death* 426)

Sayers named the novel in which Miss Climpson first appears *Unnatural Death*: a strange title considering the fact that almost all detective mysteries are motivated by an "unnatural death." But when we learn that Miss Climpson is one among thousands of "old maids, simply bursting with useful energy, forced by our stupid social system into hydros and hotels and communities and hostels and posts as companions" (427), we realize that the title may refer to the undignified end of many a spinster in Sayers's day. Of course, Sayers recognized that it was not any better for married women, foregrounding in "Are Women Human?" a problem still current today: that while a wife often has to choose between career and family, it is not necessary for a husband "to choose. He gets both" (1938, 26).

In her commitment to equal opportunities for both genders, Sayers advocates a woman's right to dress in the dominant marker of male privilege: trousers. Once again, however, she argues that women should choose their dress not merely "to copy the men" (1938, 22), an effort which merely reasserts male priority. Instead, women should choose their individual "comfort" as human beings over traditional performances of gender. In "The Human-Not-Quite-Human," she writes, "*Vir* has made the trouser his prerogative, and has invested it and the skirt with a sexual significance for physiological reasons" (1941, 38). She implies that "sexual significance" is culturally constituted, a significance reiterated and thus reinforced through the signs of gender that one performs: gender is performative.

Throughout her life, Sayers seems very aware of the performativity of gender. Several weeks after the fifteen-year-old addresses Ivy in the voice of a male lover, she writes like an infatuated schoolgirl about a male actor she has just seen on stage:

> I have fallen madly, hopelessly, desperately in love with the splendidest, handsomest, loveliest, most magnificent man in England.... And his smile!!!!!!!!! *Adorable! Ravishing!! Exquisite!!!* I have four photographs of him in my bedroom and I kiss them every night! (*Ltrs* 1. 13)

Dorothy signals that she is being self-consciously histrionic—not only through her excessive employment of exclamation points, but also when she proclaims in the next paragraph, "I doubt whether he has ever seen me!! What complicates matters further is the fact that the gentleman is already married!!!" Like the pose in Roman garb she once struck before a camera to perform feminine allure (figure 2.1), Dorothy's hyperbolic style in this letter performs, and then ironizes, adolescent female infatuation.

Dorothy had used similarly dramatic language several months earlier to describe a crush on a boy whom she and Ivy dubbed "Dull Red":

> I have got Dull-Reditis! I have been steeped in Dull Red every since you left! He leaves alas! to-morrow, but my acquaintance with him has progressed so much during this last day or two that I feel I can rest on my laurels. For it was nearly all my own cleverness. (*Ltrs* 1. 7)

Her reference to laurels and cleverness (repeated several lines later) indicates that the girl knows that she is putting on an act, manifesting the proclivity to "*pure* dramatization" that she describes in *The Mind of the Maker*: "the first literary efforts of the genuinely creative commonly deal, in a highly imitative manner, with subjects of which the infant author knows absolutely nothing, such as piracy, submarines, snake-infested swamps, or the love-affairs of romantic noblemen" (145). Sayers most certainly reflects herself in these words, having written in "My Edwardian Childhood" several years earlier that she made herself "the heroine (or, more often, the hero) of countless dramatic situations; but at all times, with a perfect realisation that I was the creator and not the subject of these fantasies" (11). Sayers sees herself controlling her performances rather than "subject" to the "fantasies" that culture constructs.

The distinction between self-consciously performing one's identity and unconsciously being subject to the performativity of discourse divides many feminists today. Cultural constructivists want to hold on to a sense of agency such that women are able to challenge the scripts of society that define the proper roles for male and female. For them, as performance theorist Marvin Carlson notes, "physical performance" offers "a possibility for women to escape" the conventional codes of society.[14] Sayers might easily fall into their camp. Aware of the gendered scripts that she enacted as a child, she subverted them as an adult by spectacularizing them in her physical performances. Susan Leonardi describes how Sayers often acted in ways traditionally considered "male": she "smoked cigars, said outrageous things, tyrannized, only half in jest, her fellow students." But other times she vamped femininity to an extreme. As one of her Somerville peers records, she appeared one morning before class "wearing a three-inch wide scarlet riband round her head and in her ears a really remarkable pair of ear-rings; a scarlet and green parrot in a gilt cage pendant almost to each shoulder."[15] Later, as a professional writer, she sometimes dressed in the "costume" of the male, causing biographers to describe her "mannish" apparel; other times she wore "feminine" markers to excess, as though to exploit them, seen with her histrionic silver wig. Though the latter was employed to cover up hair loss, a problem which recurred with stress in her life, Sayers could have worn something traditionally conservative. Instead, she chose a wig that made her look like a stage actress playing Marie Antoinette—a role reminiscent of the wigged figures of Athos and Shylock she played in her adolescence. As Mary Russo suggests, "To put on femininity with a vengeance suggests the power of taking it off," thus exposing the performative nature of gender.[16]

Many poststructuralist feminists are much less sanguine about subverting cultural constructions, for they see these constructions as inextricably embedded in discourse: the very language we use shapes our subjectivity and hence even the way we rebel against cultural codes. As theater critic Alisa Solomon wisely notes, "[G]ender isn't *merely* performance. We may self-consciously choose our clothing for effect, but few people wake up each morning and decide which gender to put on for the day. It's too simple to say that gender is *all* role-playing."[17] For feminists like Solomon, gender identity is subject to—and hence subjected by—a symbolic system that is difficult to escape. Sayers's employment of male pronouns provides a good example. In *The Mind of the Maker* she repeatedly refers to the writer as "he," even though her understanding of authorship comes from her own experiences as a (woman) writer. She uses the male pronoun with impunity, of course, because that is the way discourse performs (her) during that time. A jarring instance occurs in *Unnatural Death* when Lord Peter Wimsey discovers Miss Climpson in Mrs. Forrest's apartment: "The body of a woman lay limply on the bed. . . . But the blood was running freely, and Wimsey could have shouted for joy at the sight. Dead men do not bleed" (563).

Sayers's employment of the word "men" to describe a woman's body demonstrates how writing often performs beyond the author's control. Taught in grammar school that "men" is the generic word to describe all humans, many women alive today can attest to subtle ways that the generic male pronoun reinforced male privilege as they were growing up; just as the pronoun "he" takes the action in a sentence, so "men" take the active roles in society. However, current eschewing of the generic male pronoun indicates a change in the discursive practices of society; its usage has been slowly destabilized as the cultural assumptions it reinforced have been questioned. And current writing performances, allowing "she" to initiate action as the subject of a sentence, contribute to our sense that a "she" might initiate the action in multiple societal roles. Thus discourse, even as it changes, is still complicit with human performances.[18]

Though Sayers mimes the practice of using the word "men" for all humans, as did most people of her day, she was nevertheless very aware of how language perpetuates problematic gender performances. In "The Human-Not-Quite-Human," she travesties discursive practices by substituting male terms for normal female constructions, in order to expose how women are taught to think about themselves as not quite human:

> Probably no man has ever troubled to imagine how strange his life would appear to himself if it were unrelentingly assessed in terms of his maleness. . . . If he were vexed by continual advice how to add a rough male touch to his typing, how to be learned without losing his masculine appeal, how to combine chemical research with seduction, how to play bridge without incurring the suspicion of impotence. (1941, 39–40)

Then, in a gesture very similar to that made by Virginia Woolf in *A Room of One's Own* (1929), Sayers gives parodic book titles to illustrate how women are repeatedly inscribed as "other":

> People would write books called, "History of the Male," or "Males of the Bible," or "The Psychology of the Male," and he would be regaled daily with headlines, such as "Gentleman-Doctor's Discovery," "Male-Secretary Wins Calcutta Sweep," "Men-Artists at the Academy." (40–41)

In perhaps her most vicious parody of discursive denigrations of women, she writes, "and when he had succeeded in capturing a mate, his name would be taken from him, and society would present him with a special title to proclaim his achievement" (40). Significantly, Sayers did not allow her name to be taken from her, a prerogative allowed female writers in her time. Only occasionally did she use her married name, Dorothy L. Fleming or D.L.F., usually in letters to her son.

Some may wonder why Sayers was sensitive to certain ways language "subjects" identity and not to others; indeed, this is the poststructuralist conundrum. During the first two-thirds of the twentieth century, structuralists argued that thought is itself created by the conventional, and hence arbitrary, signifiers of language: "it is language which speaks, not the author," as Roland Barthes put it.[19] Structuralists, however, questioned all language but their own, as though their thoughts about discourse were innocent of linguistic coding. *Post*structuralists therefore argued that it is impossible to step outside of our language to fully assess its power over us, for to think about language is necessarily to employ language. Thus, while we may comment on the blindness of others, as did Sayers in "The Human-Not-Quite-Human," we are inevitably blind to our own linguistic myopias, as was Sayers to her employment of gendered pronouns.[20]

The performativity of language might even explain Sayers's refusal to consider herself a "feminist." In 1936, two years before she delivered "Are Women Human?" to a women's society, Sayers wrote her agent,

> I have a foolish complex against allying myself publicly with anything labelled feminist. . . . the more clamour we make about "the women's point of view", the more we rub it into people that the women's point of view is different, and frankly I do not think it is. (*Ltrs* 1. 391)

While she thinks she's being unusual, "kicking down the ladder I climbed up by," she actually reflects anti-feminist attitudes among many professional females during the 1920s and 1930s. For these women, "feminism usually took the form of independently written criticism or independently organized cultural activity rather than direct involvement with activist groups."[21] The statement of a Greenwich Village literary group sounds very much like their contemporary female writer, Dorothy L. Sayers: "We're interested in people now—not men and women."[22] Sayers may think she's being unique when she is actually performing the already written.

Despite poststructuralism's assumption that thought is *always* already written, thus limiting the possibility of human agency, feminists in the last quarter of the twentieth century were attracted to the "deconstruction" of Jacques Derrida. His questioning of language focused, in part, upon the binaries that have molded Western discourse—male/female, center/margin, reason/intuition, objective/subjective, truth/fiction, reality/appearance, thought/language—where the first term has traditionally operated as the origin or foundation for the second term. Derrida did not merely invert these polarities to create a new counter-system; he destabilized them—"deconstructed" them—to show that the privileging of one term over the other perpetuates inequities.[23]

In "The Human-Not-Quite-Human," Sayers anticipates Derrida when she notes how "male" is the privileged term in the binary male/female, such that the word "human" designates the male gender: "Man is always dealt with as both *Homo* and *Vir*, but Woman only as *Femina*" (1941, 37). Furthermore, like poststructuralist feminists, she recognizes that to invert the binary, such that "female" becomes the privileged term, merely reinscribes a system by which the self is identified through what it rejects as "other." She responds to a suggestion that she lead a Women's Crusade with a repudiation of any sort of abjection:

> I think it is a mistake to talk about "replacing" the masculine conception of this and that by a feminine one. That would be mere aggression

> on the part of the female! What is wanted is a merely "human" approach to every question, with neither sex trying to force the other under any sort of domination. (*Ltrs* 2. 368–69)

This may explain why she began "Are Women Human?" with the statement that begins this chapter: "I was not sure I wanted to 'identify myself,' as the phrase goes, with feminism," since the "time for 'feminism,' in the old-fashioned sense of the word, had gone past" (1938, 17). The "old-fashioned" feminism to which she alludes may be the turn of the century "first-wave" movement, whose members proclaimed female superiority, as when Frances Swiney published the statement, in 1906, that "man, on a lower plane, is undeveloped woman."[24] Rather than this inversion of the hierarchy, Sayers wants to take it apart, to "deconstruct" it, as it were: " 'Accepted as a human being!'—yes; not as an inferior class and not, I beg and pray all feminists, as a superior class" (1938, 33). Hence, she has Harriet, while on her honeymoon, object to the oft-heard wifely boast, "My husband would do *anything* for me," asserting that "No human being ought to have such power over another."[25] And though Harriet was speaking of female power, Sayers obviously indicts male power as well.

Today, some feminists respond to the power of male discourse by positing an essentialism to female experience, asserting that, if language constrains and polices behavior according to patriarchal paradigms, one must appeal to a stage of the self before it was molded by the performativity of language. Many of these feminists have been influenced by Jacques Lacan's poststructuralist interpretation of Freud. For Lacan, a child's Oedipal entry into self-awareness was an entry into the "symbolic" realm of language, what he calls the "Law of the Father." Preceding this stage, the child still participated in the "imaginary" realm of identification with the mother. Feminist theorist Julia Kristeva picked up on this division, renaming Lacan's "imaginary" the "semiotic," which she identified with intuitive, nonlinear, sensuous communication with the mother. Kristeva valorized the semiotic as a feminist language toward which males as well as females could aim in an effort to subvert the Law of the Father.

Other essentialist feminists celebrate ways the body itself might engage in non-patriarchal communication—through the mother–daughter relation (as in Nancy Chodorow), through a signifying practice that "writes" the female body (as in Hélène Cixous), or through the two lips that move between a woman's legs (as in Luce Irigaray). Such mystifying of the female body, however, bothers other feminist critics. Some question how the body can be identified apart from discourse, while others assert that essentializing the body can perpetuate traditional gender performances.[26]

Margaret Homans addresses the relationship between the body and language in *Bearing the Word*, calling the female body "the literal" and language "the figurative." Employing Lacan's revision of Freud, she asserts that, once males have been separated from the mother due to the Oedipal incest taboo, they "view the mother as the literal, she whose absence makes language both necessary and possible," meanwhile regarding "the literal" as "valueless and the figurative valuable."[27] She then outlines various ways nineteenth-century women writers, defying "women's placement as the silent object in the traditional male view of language," sought empowerment by employing the "figurative" language of the father (33). This desire to "bear the word," as Homans puts it, might explain Sayers's proclivity to perform—to actually speak—male roles. When the performances were written, Sayers experienced the control of the "figurative" traditionally aligned with male power. As with Simon de Beauvoir, Sayers's near contemporary, "writing constituted an assertion of will, a transcendence of environment, a renunciation of those aspects of the Eternal Feminine . . . which inculcate women's silence."[28]

This "renunciation," however, comes perilously close to a repudiation of the mother, whose body represents the objectification—or "literalizing"—of the female. Therefore, women writers often engage the "figurative" realm of language to honor the "literal" body of the mother, at times invoking the mother of all mothers: the Virgin Mary, "who gives birth to and is frequently imaged carrying (thus two senses of 'bear') a child who is the Word, the embodiment of the Logos" (Homans 30). Sayers, of course, employs this charged image in her play cycle about the life of Jesus, *The Man Born to Be King*, calling Mary "fact" and God "truth," asserting that literal body (fact) and figurative meaning (truth) were united through the Incarnation of Jesus. In *The King of Sorrows* play, the Virgin states, "I, Mary, am the fact; God is the truth; but Jesus is fact and truth—he is reality. You cannot see the immortal truth till it is born in the flesh of the fact" (*King* 289). In the next scene, Sayers draws attention to the importance of Mary Magdalen's body—the "literal"—while Jesus is being crucified. Roman soldiers will not allow the two Marys and John to approach the cross until Magdala reminds them of pleasures provided by her body. Removing her veil and unpinning her red hair, Mary sings and dances for the soldiers until they allow access to Jesus. Mary's body provided means to the truth.

Even at the start of her fiction-writing career, Sayers registers a hope that the Law of the Father might cease to be the only access to truth. On the second page of her first novel, appropriately titled *Whose Body?*, she has the revelation of the murdered body come from the mouth of the mother, as the Dowager Duchess telephones her son to report the naked man found in a bathtub (8). It is as though the mother identifies with the "literal" object of the male gaze, which a murdered body always becomes. Significantly, the Dowager is not at all constrained by the "figurative," often misusing words and employing digressions in illogical, nonlinear ways. She also does not follow the prescriptive dogmas endorsed by the Law of the Father, such as class privilege, whereby, as Peter notes, "Dad always hated self-made people and wouldn't have 'em at Denver. And old Gerald [Peter's brother] keeps up the tradition." Peter reflects on his mother's difference, asserting that he trusts her "judgment a good deal" (126): a judgment based not on the "figurative"—the Law of the Father—but on the "literal" as she cares for the bodies of others. In *Whose Body?* she comforts Thipps' deaf mother and Levy's widow; in the following novel, *Clouds of Witness*, she understands Mary's body, recognizing her deceptions that have been psychologically and physically analyzed—incorrectly—by male doctors. The Dowager tells the doctors, "You are all perfectly ridiculous, and no more fit to take care of yourselves than a lot of babies—not but what there are plenty of poor little things in the slums that look after whole families and show more sense than the lot of you put together" (115). Her non sequitur about the little things in the slums indexes how her empathy for literal bodies overrides her figurative precision. Peter shares this trait, though to a less obvious extreme. While his creator spectacularizes the performance of the "literal" by the way she dresses her body, Peter spectacularizes the performance of the "figurative" by the way he dresses his language, often subverting serious conversation by throwing in garbled bits of songs, clichés, and verse.

Peter even solves cases not through "elementary, my dear Watson" figurative logic, but through the intuitive leaps characteristic of his mother. Significantly, in *Clouds of Witness*, Sayers self-consciously aligns detection with "mother-wit" as the Dowager Duchess describes her "detective instinct": "My dear child, you can give it a long name if you like, but I'm an old-fashioned woman and I call it mother-wit, and it's so rare for a man to have it that if he does you write a book about him and call him Sherlock Holmes" (117). Peter, then, solves crimes not by copying the "figurative" Law of the Father, but often through the intuitions modeled by his mother's wit. Take, for example, his solution to the crime in *Whose Body?*:

> It happened suddenly, surely, as unmistakably, as sunrise. He remembered—not one thing, nor another thing, not a logical succession of things, but everything—the whole thing, perfect, complete, in all its dimensions as it were and instantaneously. . . . He no longer needed to reason about it, or even to think about it. He knew it. (127)

Significantly, this statement of illumination, like a sunrise in the brain, follows less than a page after Peter pronounces, "I trust Mother's judgment a good deal." Furthermore, once he catches the murderer, he shares his mother's empathy for the literal body, experiencing a nervous attack—a relapse into shell shock from the Great War—when he must send the criminal to his death.

Biographers sometimes relate Peter's relapses of shell shock to the experiences of Sayers's husband, who was gassed during the war; however, Peter's first attack, in *Whose Body?* (130–31), was written before Sayers had met Mac Fleming.[29] In contrast, six years before *Whose Body?*, Sayers writes an Oxford friend about a cousin and uncle visiting from the front: "I'm rather glad they're gone, because I do so hate everlasting war-talk, and I'm always in terror of Mother getting another of her nervous attacks. . . . [T]hey are so terrifying, because she loses control over her speech and limbs" (*Ltrs* 1. 110). Sayers gives Peter not only a mother-wit, but a body that experiences attacks like those of her own mother. Even in her last published novel, *Busman's Honeymoon*, Sayers inscribes Peter with the "literal" experience of her mother, reinforced by the fact that it is the Dowager Duchess who explains to Harriet the origin of her son's nervous attacks.

Identification with the "literal" realm of the mother, however, can limit a writer's sense of agency, as Homans makes clear in *Bearing the Word*. This may explain why Sayers decided she needed a more figuratively functional female than the Dowager Duchess to help Peter in his sleuthing. Hence, for her third novel, *Unnatural Death*, she created the perspicacious Miss Climpson, modeling her after another woman in her family, Aunt Gertrude Sayers.[30] Two novels later, *The Documents in the Case*, she leaves Peter out entirely, supplying the protagonist with "a modern woman and a successful novelist" (18): an astute female who can challenge her fiancé's sexist assumptions. As Catherine Kenney suggests, it was through this couple that Sayers "discovered the direction the later Wimsey books were to take."[31]

In *Strong Poison*, published soon after *The Documents in the Case*, Sayers returns to Peter, supplying him with Harriet Vane, another modern woman and successful novelist. However, while his mother advises him and Miss Climpson helps him, Harriet Vane changes him: "Like the poisoned Athulf in *Fool's Tragedy*, he could have cried, 'Oh, I am changing, changing, fearfully changing'" (53). By paralleling Peter with the poisoned Athulf, Sayers hints that the "strong poison" signalled by her title is not just that which kills Harriet's former lover. Significantly, Harriet had bought the arsenic that implicates her in the performance of murder in order to do research for one of her writing performances. The poison that aligns her with the "literal"—a dead body—is actually a sign of her figurative abilities. Harriet's poison is her pen, which gives her power over language that kills cultural clichés about the material female.[32]

Nonetheless, other than the agency she exercises by refusing Peter's proposals of marriage, Harriet is imprisoned in *Strong Poison*—literally as well as figuratively—by a patriarchal system that judges woman by her body. Because she had been living with the murdered victim without benefit of marriage, Harriet is considered guilty by sexual association. As Susan Rowland puts it, "In Sayers's *Strong Poison*, Harriet functions as the colonised outsider whose unconventional sexual behaviour condemns her as other in the eyes of England's establishment."[33]

In the next novel in which she appears, *Have His Carcase*, Harriet is once again aligned with the body when she discovers the murdered corpse. Significantly, when she meets the murderer, he treats her merely as a body: the sexualized object of his gaze. Not only does he try to grope her flesh, but he regards "women's rights" as a type of "mania" comparable to "crystal-gazing" (286, 235). Peter, in contrast, treats Harriet as an intellectual equal, building upon her figurative discoveries to help her solve the case.

He does the same, of course, in *Gaudy Night*, only this time Sayers makes it even clearer that Peter's presence does not signal female inadequacy. Similar to her tactic in *Have His Carcase*, Sayers makes anti-feminist assumptions a manifestation of the criminal mind. When Peter exposes Annie as the perpetrator of the crimes against Shrewsbury College, the scout lashes out at the dons, who pride themselves in their figurative abilities as scholars: "You can't do anything for yourselves. Even you, you silly old hags—you had to get a man to do your work for you" (*Gaudy* 487). Ironically, those critics who accuse Sayers of sexism in *Gaudy Night* sound much like Annie. SueEllen Campbell states, "Despite the novel's explicit endorsement of the intellectual independence and ability of women, . . . Sayers still allows the one man involved to have the final authority." Margot

Peters and Agate Krouse go one step further to say that Peter "never indicates that the successes for which he is repeatedly congratulated rest on the work of women; nor does Sayers remind us of that fact on appropriate occasions."[34]

Perhaps Sayers is subtler than her critics realize. In *Gaudy Night*, she spectacularizes two kinds of gendered writing performances. One is staged by the women dons, for whom the knowledge that someone had "loved and sinned and suffered and escaped death was of far less ultimate moment than a single footnote in a dim academic journal" (18). The other performance is by Annie, who proclaims a commitment to her "own flesh and blood" (488): the realm of the literal. Believing that the professional commitment of the dons to the intellectual realm of the figurative—the Law of the Father—has killed the father of her children, Annie sabotages their scholarship and vandalizes the repository of the figurative: the library. Annie's destructive impulses are not deconstructive, however; assuming that the dons have inverted, and hence perverted the proper relationship of man and woman, she seeks to perpetuate the binaries of conventional culture, where mind is to body as male is to female: "A woman's job is to look after a husband and children.... The whole bunch of you together haven't flesh and blood enough to make you fit for a man" (486, 487). Following a gendered script of her day, Annie asserts that the dons's entry into the male realm of the figurative has eviscerated their literal bodies. Annie thus takes a stand against the deconstruction of gendered binaries.

As well as problematizing the traditional binaries of Western culture, deconstruction, as an interpretive practice, exposes how a person's intentions are often undermined by the language employed to express them. Annie, of course, intends to desecrate the figurative prowess of the Shrewsbury dons, believing that the literal—"flesh and blood"—is the realm of authentic female performance. Ironically, her vandalism performs almost entirely in the realm of the figurative: a nasty drawing, a dirty postcard, a libelous note, a vulgar epithet, graffiti covered walls, letters of harassment. Even when a body ominously appears, a don hanging from the chapel ceiling, the object turns out to be not literal but figurative: an effigy, impaled with a knife stuck through a sheet of Virgilian verse. The mysterious creator of all these texts therefore becomes known by the dons as the "poison pen"; she becomes identified with the representative implement of the figurative.

In their groundbreaking work of feminist criticism, *The Madwoman in the Attic: The Woman Writer and the Nineteenth-Century Literary Imagination*, Sandra Gilbert and Susan Gubar establish the "pen" as a metaphoric penis: the implement of figurative empowerment traditionally acceded to males. By making the bearer of the poison pen at Shrewsbury College a widow who believes that woman's role should be defined by her material body, Sayers deconstructs the binary Annie sets up between the literal and the figurative. Annie's writing performances, which disrupt an entire college without anyone ever witnessing her body, demonstrate her dependence upon that which she repudiates. In a brilliant move, Sayers applies this deconstruction to the larger workings of the novel, refusing to include the conventional impetus for the sleuthing of detective fiction: a dead body, the flesh and blood of the literal.[35] In *Gaudy Night*, a pen, not a body, is poisoned.

The problematics of gendered performances surface also in Harriet's worries about marriage and how it might affect her writing performances:

> Could there ever be any alliance between the intellect and the flesh? . . . Experience perhaps had a formula to get over this difficulty: one kept the bitter, tormenting brain on one side of the wall and the languorous sweet body on the other, and never let them meet. (*Gaudy* 458)

In order to deconstruct this wall-like cultural construction—the binary division between male mind and female body—Sayers aligns Peter, in *Gaudy Night*, with the realm of the literal. Donald Marshall explains the significance of Peter's arrival at Shrewsbury College by quoting Harriet's response:

> She had not "considered him primarily as a male animal, or calculated the promise implicit in the veiled eyes, the long, flexible mouth, the curiously vital hands." She had felt in him no "domination but that of intellect." But as he walks across the Shrewsbury Quad, under the gaze of the Senior Common Room, she suddenly sees him with "the eyes of women who had seen him before they know him," that is, registered him sensually and not mentally.[36]

Marshall establishes Peter as the object of "the gaze."

In her famous essay, "Visual Pleasure and Narrative Cinema," Laura Mulvey discusses "the gaze," explaining how "to-be-looked-at-ness" is traditionally aligned with females being observed by males.[37] Sayers subverts this tradition when she has Peter become the object of the female gaze, not only within Shrewsbury College, but also as Harriet studies the intricacies of his face as they float down the Thames. Peter becomes identified by his body, thus performing the conventional role of woman. According to Carolyn Heilbrun, "Only

Colette has also accomplished this rare feat of brilliantly rendering the male as sex object."[38]

In *Gaudy Night*, then, Sayers constructs the binaries of brain/body, intellect/flesh, head/heart, figurative/literal in order to show Peter and Harriet rising above them. Significantly, she has Harriet surmount the wall between tormenting brain and sweet body the moment she mounts to the roof of the Radcliffe Camera. Peter, having found her studying in the Camera, suggests the ascent, at which point we are told "Harriet put down her pen" (*Gaudy* 493–94). Momentarily letting go of her symbolic means to the figurative, that traditionally male realm which enables her to escape the exigencies of the female body, Harriet is finally able to discuss her relationship with Peter. When he subsequently leaves the Camera, Peter functions, once again, as object of her gaze: "Harriet was left to survey the kingdom of the mind, glittering from Merton to Bodley, from Carfax to Magdalen Tower. But her eyes were on one slight figure that crossed the cobbled Square" (498). The body of this figure, she realizes, will allow her to engage in the figurative body of her writing performances.

As Harriet gazes out from Radcliffe Camera, she sees Oxford as "the kingdom of the mind": a kingdom that should be ruled without regard to gender distinctions. To symbolize the effacement of these distinctions, Sayers draws attention to the black academic gowns worn by male and female alike. During his first visit to Shrewsbury College, Peter mistakenly confuses his gown with Harriet's:

> "Bless the man, if he hasn't taken my gown instead of his own! Oh, well, it doesn't matter. We're much of a height and mine's pretty wide on the shoulders, so it's exactly the same thing."
>
> And then it struck her as strange that it should be the same thing. (311)

Sayers, it seems, expects us to be struck as well that the sign of education can be an androgynous signifier.

In "The Human-Not-Quite-Human" Sayers asserts that the human brain is an "Androgyne," able to "mate indifferently with male or female and beget offspring upon itself" (1941, 44). Once again she sounds like Virginia Woolf, who discusses in *A Room of One's Own* Coleridge's idea "that a great mind is androgynous," noting that "It is when this fusion takes place that the mind is fully fertilised and uses all its faculties" (102). Both women use sexual metaphors—"mate," "beget," and "fertilised"—to discuss mental processes; for them, bearing the word necessitates an interpenetration of the literal and the figurative in both males and females.[39] In *Gaudy*

Night, this interpenetration is promised through Harriet and Peter's marriage, where the pen in both its literal and figurative manifestations will be engaged, as Sayers makes clear in "Talboys," a short story in which Harriet still bears the word after bearing three children.[40]

In light of Sayers's androgynous purposes, the closing paragraph of *Gaudy Night* makes more sense:

> The Proctor, stumping grimly past with averted eyes, reflected that Oxford was losing all sense of dignity. But what could he do? If Senior Members of the University chose to stand—in their gowns, too!—closely and passionately embracing in New College Lane right under the Warden's windows, he was powerless to prevent it. He primly settled his white bands and went upon his walk unheeded; and no hand plucked his velvet sleeve. (501)

Sayers eliminates sexual distinctions with the word "Members." Wearing identical gowns, the figurative and literal embrace in such a way that renders gendered performances irrelevant. The Proctor, however, is coded as the masculine figurative; the object of no one's gaze, he walks unheeded.

Significantly, the androgynous kiss of Peter and Harriet follows a performance of Bach where they sit side by side in their gender-neutral gowns:

> Peter, [Harriet] felt sure, could hear the whole intricate pattern, every part separately and simultaneously, each independent and equal, separate but inseparable, moving over and under and through, ravishing heart and mind together. (499)

The performance of Bach's "polyphonic" music becomes a metaphor for Peter and Harriet's relationship, which deconstructs the hierarchy of male over female, intellect over body, center over margin. The destabilizing of such binaries, however, is not easy. As one of the dons, Miss de Vine, tells Harriet, "If you ever find any kind of repose with [Peter], it can only be the repose of very delicate balance" (491). Miss de Vine speaks from experience; it was her commitment to the realm of the figurative that led to the suicide of Annie's husband, for she took no interest in the literal needs of his body after she accused him of plagiarism, thus depriving him of a job. Miss Lydgate, she believes, would have been more balanced: she "would have done what I did in the first place; but she would have made it her business to see what became of that unhappy man and his wife" (489–90). Miss Lydgate, in other words, would have operated from an androgynous perspective, committed to both integrity in the realm of the figurative and compassion for the literal needs of the body. As Carolyn Heilbrun notes in *Toward a Recognition of Androgyny*, the delicate balance of male and female characteristics within one person "seeks to liberate the individual from the confines of the appropriate" gender performances (x).[41]

Commitment to a delicate balance between the figurative and the literal seems to evaporate in the next Harriet–Peter vehicle, *Busman's Honeymoon*. Harriet, on her honeymoon with Peter, reflects the passivity accorded married women in her day. When a body is discovered in the cellar of their honeymoon cottage, it is the men who search for the figuration of clues, while "Harriet, relegated to the woman's role of silence and waiting, went into the kitchen to help with the sandwiches" (113). Even at the end of the novel, Harriet sits in "gracious silence" (369), waiting for Peter to share the despair over *his* role: as an amateur detective responsible for the death of the criminal. *Busman's Honeymoon* thus illustrates an irony often manifest in women's detective fiction: "The male detective operates outside society's conventions because that is what male heroes do; the woman detective should not be a detective *because* she operates outside society's conventions, and that is not what female heroes should do."[42] However, there may be more going on in *Busman's Honeymoon* than meets the eye, especially in light of the unusual circumstances of its inscription.

Originally written for the stage, *Honeymoon* was suggested and encouraged by Muriel St. Clare Byrne, the friend who collaborated with Sayers on the script.[43] As they were waiting for the play to find a producer, Sayers shaped the story into novel form. Although the detective plot, though embellished, is basically the same in both genres, Sayers adds to the novel a framing device: a "Prothalamion" at its start and an "Epithalamion" at its end. The latter word has been used since Classical Greek culture to describe a poem or song praising a bride and groom upon their nuptials; "Prothalamion" is a word Edmund Spenser invented in 1596 to title a poem celebrating a double wedding, since he had already used "Epithalamion" to honor his own marriage. Sayers employs these age-old terms, I would argue, to signal that she wants us to think about the institution of marriage itself, and how it scripts gendered performances.

The Prothalamion, as a collection of documents written by several different people, is reminiscent of *The Documents in the Case*, the novel that directly preceded Harriet's entrance into the Wimsey canon. In the *Honeymoon* novel, various letters about the Wimsey–Vane wedding, plus the Dowager's journal entries, provide multiple perspectives on the event, especially of Peter's demeanor during the ceremony. Peter's sister-in-law, the Duchess of Denver, writes that her brother-in-law looked "as white as a sheet; I thought he was going to be sick. Probably he was realising what he had let himself in for" (5). The Dowager Duchess, however, reporting on the same white countenance, interprets it as tremulous excitement—"like the day he had his first watch, and could hardly bear himself for fear it would come to pieces in his hands or turn out not to be real" (25). Meanwhile, the Dean of Shrewsbury College describes Peter as "the only bridegroom I have ever seen who looked as though he knew exactly what he was doing and meant to do it" (9). Sayers implies that culturally scripted performances, marriage included, are open to interpretation, and that perhaps we should not jump to conclusions about the apparent conventionalities of the Wimsey marriage in the main body of the text.

Sayers even provides several references to actual literary texts in the Prothalamion to underscore the issue of interpretation. Four times the Dowager refers to *The Stars Look Down*, a 1935 novel that presents multiple interpretations of people and events.[44] The Dowager also notes that when Peter presented a mink cloak to Harriet, the latter responded, "'Oh, Mr. Rochester!'—in fun, and meaning Jane Eyre" (23). This allusion reminds us of the subversion of convention that appears in Charlotte Brontë's famous novel, where the aristocratic Rochester, like Peter, marries a commoner, having redefined what it means to be "equal." However, to be genuinely equal to Jane Eyre, as Elaine Showalter notes, Rochester must have "a limited experience of dependency, frustration, and powerlessness—in short, of womanhood."[45] Brontë thus has Rochester lose a hand and his eyesight before he marries Jane Eyre, such that he must rely on her strength in the midst of his powerlessness.

Sayers signals a similar powerlessness in Peter at the end of her novel. The Epithalamion opens with a focus upon the hero of Harriet's detective fiction, Robert Templeton, who "would unmask his murderer with a flourish of *panache* in the last chapter and retire promptly from the stage amid a thunder of applause" (335). Sayers's references to the "last chapter" and "the stage" may allude to the fact that the preceding, or "last," chapter of the novel parallels the staged version of *Honeymoon*, where Peter and Harriet retire from the text with the flourish of a kiss. However, as the narrator goes on to note, "real life" differs considerably from the *panache* of the stage—as though to foreground the problematic conventionality of many writing performances. As the Epithalamion continues, the Dowager explains to Harriet the origin of Peter's shell shock, which, as we have seen, resurfaces whenever Peter feels responsible for the death of others, as he did in the trenches of the Great War. Thus, while the staged *Honeymoon* presents Harriet in a more conventional, passive role, Sayers's novelistic addition makes Peter dependent on Harriet. Wanting to write a letter to Miss Climpson, Peter sits "blankly, pen in hand," until he finally hands the pen over to Harriet, asking her to write the letter for him.

At the close of the novel, Peter and Harriet embrace as equals, but not in the form of the conventional kiss that ends the *Honeymoon* play. Instead, the image is of a crying baby seeking a return to the body of the mother: "Quite suddenly, he said, 'Oh, damn!' and began to cry.... So she held him, crouched at her knees, against her breast, huddling his head in her arms" (369). Significantly, Peter's last spoken line in this last novel Sayers was to publish—"Oh, damn!"—is the same as his first spoken line in her first novel, *Whose Body?* Peter's initial employment of the phrase indexes his reliance on the realm of the figurative: he has left behind a catalogue of books. In *Honeymoon*, of course, the phrase indicates his dependence upon the literal—the female body. As with the polyphonic music the couple enjoy while sitting in their matching academic gowns at the end of *Gaudy Night*, in the Epithalamion Peter and Harriet seem "separate but inseparable," their identities "moving over and under and through" each other, "ravishing heart and mind together" (*Gaudy* 499).

Six years before Sayers wrote the polyphonic into *Gaudy Night*, Mikhail Bakhtin published his monograph on polyphony in the novel: *Problems of Dostoevsky's Creative Work*. The polyphonic novel, for Bakhtin, presents multiple subject positions without needing to establish one voice as authoritative or to harmonize the diversity of positions. As Peter Wimsey puts it, "anybody could have the harmony if they would leave us the counterpoint... I like my music polyphonic" (*Gaudy* 499). Sayers's affirmation of the "polyphonic," like her desire for a "delicate balance" and an androgynous mind, points to how she grapples with her own multiple subject positions as they become staged in her writing performances.

In *The Mind of the Maker* Sayers decries how "we are continually tempted to confine the mind of the writer to its expression within his creation," trying "to identify him with this or that part of his works, as though it contained his whole mind" (50). She thus recognizes, with Bakhtin, that "the idea" within the author's mind "is inter-individual and inter-subjective—the realm of its existence is not individual consciousness but dialogic communion *between* consciousnesses."[46] Such communion between consciousnesses reaches its full when Harriet agrees to wed Peter. Her earlier resistance to Peter's marriage proposals within *Strong Poison*, *Have His Carcase*, and much of *Gaudy Night*, might therefore be read as a resistance to what Bakhtin calls "monologism," of the submission of one consciousness to another. Indeed, Harriet refuses Peter because she worries that her gratitude for actions that saved her life will prevent her from being completely herself around Peter, that she will not be able to resist his ideas when they differ from her own. She changes her mind when she realizes that he desires a "delicate balance" among polyphonic positions, wherein their "specific world views, each characterized by its own objects, meanings and values," might "mutually supplement one another, contradict one another and [become] interrelated ideologically."[47]

When biographers identify Sayers with Harriet Vane, then, they limit the androgynous mind of the maker. Sayers endows different characters—some male, some female—with various elements of her polyphonic subjectivity, later noting that "the writer has 'favorite' characters, which seem to embody more of or more important parts of his personality than the rest" (*Mind* 52). Peter's servant, Bunter, for example, is an avid photographer, as was Sayers, who at private school won an award for her photographic prowess and who writes in a 1925 letter, "I take rather fine photographs" (*Ltrs* 1. 223).[48] The aristocratic snobbery that bothers many readers, wherein Peter orders Bunter about with impunity, is thus mitigated somewhat by Peter's respect for Bunter's admirable qualities, the most important being his commitment to work well done. Indeed, Bunter respectfully challenges his employer when the integrity of their detection is on the line, repeatedly eliciting from Peter the statement, "You are always right, Bunter." The same way Bunter regards his work in service to Peter, Sayers regards her work as Bunter's creator, explaining in *The Mind of the Maker* that a creator's business is to serve her creation: "but to serve it he must love it. If he does so, he will realize that its service is perfect freedom" (*Mind* 66). Bunter's service to Peter is also perfect freedom.

Like Bunter, Chief Inspector Charles Parker is committed to the integrity of his work. In contrast to Peter, who has "nothing much in the way of religion" (*Gaudy* 496), Parker operates from a Christian perspective, expressing it, like Sayers, as an integral part of one's life, rather than with evangelistic zeal or polemical fervor. Parker ends up marrying Peter Wimsey's sister, Mary, who also manifests an element of Sayers's psyche. In *Clouds of Witness*, Mary is madly in love with a Russian socialist, Goyles, with whom she plans to elope. Goyles, however, turns out to be more concerned with protecting his own interests than guarding those of Mary and, in great pain, she breaks off the relationship. There is more than a mild resemblance here to the relationship Sayers developed in 1921 with the Russian-born John Cournos, whom she desperately wanted to marry but who told her he didn't believe in marriage. Because Sayers was in the middle of writing *Clouds of Witness* when she heard that John had married an American detective fiction writer (summer 1924), it is highly likely that Mary's disappointment in her Russian lover expresses Sayers's depression over Cournos. Indeed, both men fail to live by their words, Cournos about marriage and Goyles according to his "public speech[es] calling for decisive action."[49]

Even Margaret Harrison, the adulterous wife in *The Documents in the Case*, though described by Sayers as a "dreadful person" (*Ltrs* 1. 302), manifests a subject position with which Sayers identified. The characterization given Margaret by Munting, her lover's flatmate, anticipates what Sayers, several years later, will say about herself. Munting writes to his fiancée about Margaret's "power of *dramatisation*":

> If ever she found anyone to take one of her *impersonations* seriously, she would probably be able to live very brilliantly and successfully in that *character* for...long enough to make an impressive *drama* of it. (*Doc* 31, emphasis mine)

We are reminded of "My Edwardian Childhood," written two years after *The Documents in the Case*, where Sayers says, "I...have at all periods of my life continued to *dramatise* myself, into a great number of egotistical *impersonations* of a very common type" (11, emphasis mine). Furthermore, the way Harrison and his son treat Margaret anticipates the way adults treat Katharine in *Cat o' Mary*: her "manifestations of extravagant feeling were snubbed and discouraged...in the name of common sense or good form" (*Cat* 107). As Margaret tells Paul Harrison, "You have just his way of snubbing and repressing people. You don't seem to understand that everybody can't keep their feelings bottled up as you do" (*Doc* 152). Those feelings come pouring out in Margaret's letters to her illicit lover: "Darling, I love you so much I don't know what to do with myself. I have kissed the paper twenty times where your dear, darling name is" (103–04). Sayers, here, reinscribes her fifteen-year-old writing performances, when she pours out her soul to Ivy: "Oh, love, love! To see thee again! To hold thee once more to my heart! I laid thy letter there—but I would embrace thine own dear self."[50] It would seem that, when Sayers began to create Margaret Harrison, she reconstructed the subject position of her adolescent performances in order to script the artifice of infatuated love.[51]

When Sayers left her detective fiction behind for the next stage of her career, she created characters for the stage that also voice positions with which she identified. Most notable is Janet Reed,[52] the protagonist of *Love All*, which was first produced in 1940. Alzina Stone Dale, in her introduction to the play, makes a good case that Reed, a playwright, expresses more of Sayers's self than does Harriet Vane. It therefore becomes significant that *Love All* refers not only to performances written for the stage, but also to the problematic performance of gender.

Love All presents a gendered double standard through Godfrey, an adulterous novelist who is shocked to learn that both his wife and his mistress take their professional work as seriously as do males. He describes as "indecent, ... ridiculous and disgusting" the fact that the two women have established a friendship based upon a mutual respect for writing performances: the wife, Janet, writes performances as a playwright, and the mistress, Lydia, performs writing as an actress.[53] It is appropriate that Lydia, trained to perform scripts, naively pronounces a socially constituted cliché: that behind every great man one finds a woman sustaining him. Janet, however, having come into possession of the pen by superceding her husband as a successful author, rewrites the script, telling Lydia, "And every great woman has had some man or other in front of her, tripping her up" (185). Godfrey, in desperation, proposes to his secretary, who, in her paid position, has performed the culturally constructed code that best serves his gendered interests: "Only from you, Mary, have I ever received the unwearying devotion—the comfort, the help, the inspiration that a man looks for in his wife. Mary, will you marry me?" (200–01). To the audience's great satisfaction, she refuses to act in Godfrey's theater of representation, defying his gendered script by turning him down.

While *Love All* presents the performance of gender in comic form, *The Zeal of Thy House*, written the year before, expresses it as semi-tragic Romance. Here, the protagonist, who parallels Sayers in his commitment to work well done, is male: the twelfth-century architect William of Sens. Hired to rebuild the burned out choir of Canterbury Cathedral, William suffers a disfiguring fall—both literally and figuratively—due to his pride. Sayers encodes as part of his arrogance the way he treats the woman who loves him: Ursula, a wealthy widow and patron of the building project. Their first conversation focuses on gender performances, with William pronouncing,

> What use have women for the dreams of a man
> Save to destroy them? ...

...Women create
Passively, borne on a wind of lust, for a whim,
At the caprice of a man, in a smile, in a spasm
Of the flesh; we, with the will, with the blood, with the brain,
All the desire of the soul, the intent of the mind. (*Zeal* 51)

William piles up, like stone upon stone, the binary parallels of male/female, mind/body, active/passive, soul/flesh. Ursula, subject to the same cultural discourse, acknowledges the contemporary reality of William's formulations, but problematizes them at the same time:

—knowledge is given to man
And not to woman; and the glory of work
To man and not to woman. But by whom
Came either work or knowledge into the world?
Not by the man. (51)

The abilities that William celebrates as "male" were made possible when the female Eve "Seized knowledge for herself" (52). Just as Janet Reed rescripts the cultural cliché of a woman behind every man, Ursula rescripts the cliché of female culpability behind the fall. Instead of leading man to his depravity, Ursula's Eve provides him with his glory: "work," a result of the Fall. Sayers makes Ursula's Eve sound like Prometheus: she "snatched the torch/of knowledge from the jealous hand of God/So that the fire runs in man's blood for ever." And instead of being thankful, man "contrives to boast" of "work"

As his peculiar glory, and in one breath
Denies it to the woman and blames her for it,
Winning the toss both ways. (52)

Like Prometheus, woman is in chains for the freedom she gave man. Sayers thus gives a feminist interpretation of the "*felix culpa*" mentioned by the angel Raphel thirteen lines later. The "*felix culpa*," or "fortunate fall," by which God glorifies humanity through the "work" of Christ is made possible through Eve's efforts for "knowledge" and Mary's efforts in childbirth: women "bearing the word" both figuratively and literally.[54]

When William experiences his own literalized fall later in the play, woman, once again, carries the blame. Two men become so obsessed with the presence of Ursula's body in the church that they fail to inspect the rope that suspends William at the top of an arch. And, as happened with Eve, they blame the woman for the man's fall. The maimed William, however, commits the opposite sin: eschewing Ursula's body. In an expression of earnest love, Ursula offers to marry William and care for him. However, he refuses to submit to the materiality of the female, still believing in the indispensability of his figurative "work": directing the building project from his sickbed. Only when the Prior, aided by the Archangel Michael, convinces him of his hubris, does William recognize his lack of connection both to God and to woman.

Sayers's perspectives on gender thus reflect her theology: because the incarnated Jesus instantiated the union of God not with man but with woman, he manifested an unusual sensitivity to females:

> There is nothing whatever in any act or word of His that suggests any peculiar mystery, danger, excitement, or oddity about women or sex; and in His dealings with women He was completely unselfconscious, treating them quite straightforwardly as human beings with minds and souls of their own. In fact, He walked straight through all the sex-taboos as though they did not exist. Neither the Jews and pagans before Him, nor the Christians and Mohammedans and neo-pagans since have ever achieved anything like His normality; only a few exceptionally well-balanced people here and there have ever got within miles of it. (*Ltrs* 2. 402)

Because of this balance, Jesus "never patronised, or condescended, or scolded, or nagged at women for being women, or turned shy or silly or self-conscious or superior on them." He did not condone the binary wherein men, having the figurative abilities, are the "complete human beings with certain sexual functions," while women are reduced to merely the literal: "perambulating sex-organs with a distant resemblance to humanity" (*Ltrs* 3. 29). It would seem that, for Sayers, Christ's avoidance of simplistic binaries stems from the delicate balance of his fully human and fully divine natures—body and word, literal and figurative in one form.

Sayers later uses the phrase "delicate balance" in a letter about orthodox views of the Incarnation, where "The Event—the Act of God in History—is all-important, because it ties the thing to this world, and to time and place and the flesh." Unfortunately, she notes, people tend to emphasize one extreme or the other, establishing a hierarchical relationship between the physically human body of Jesus—the literal—and his figurative manifestation as a sign of God presence:

> Hence heresies and other distortions, which are *always* due to an over-simplifying emphasis of one truth at the expense of the others—the

> "nothing-but" system of interpretation which sits heavily down on one little bit; whereas the whole truth is always walking a razor-edge of delicate balance between the lop-sided exaggerations. (*Ltrs* 4. 410)

One lop-sided exaggeration that repeatedly aggravated Sayers was Gnosticism, a Christian heresy that perpetuated the Greek privileging of knowledge over body, what we might call the dominance of the figurative over the literal. Sayers, appropriating an image from Dante, warned a correspondent about the "evil way of the Gnosis," telling her to "steer clear of Knowledge with a capital K—or you'll end up with your face twisted over your backside in the *Bolgia* of the Sorcerers!" (*Ltrs* 3. 275). Domination of the figurative can warp the materiality of the body—as has happened to females, who are "forced to think of themselves in terms of sex" while concomitantly "jeered at for not being capable of thought" (*Ltrs* 3. 29).

Having been taught by her body, which performed well in both male and female dress, Sayers advocated an androgyny of the mind, believing that writing performances must include both the literal and the figurative, the material and the symbolic, the image and the idea, acting in "delicate balance"—like the *Imago Dei*. Sayers began to formulate her theology of the *Imago Dei* in *The Zeal of Thy House*, which alludes to the need for delicate balances through William's description of Gothic architecture: "A triumph of balance. . . . A delicate adjustment of interlocking stresses" (42). As we will see in the next chapter, gender is not the only cultural construction that occupied Sayers's androgynous mind; a zeal for architecture—for literal constructions—profoundly affected Sayers's writing performances.

CHAPTER 3

THE PERFORMANCE BUILDS: SAYERS'S ARCHITECTURAL IMAGINATION

. . . the creative artist does,
somehow or other,
specialize in construction.

—Sayers, *The Mind of the Maker*

The performances which most exhilarated Sayers during her years at Oxford were those of Hugh Percy Allen (1869–1946), organist at New College and conductor of the Oxford Bach Choir, which Sayers joined soon after she entered Somerville College. After attending Dr. Allen's Orchestra Concert in October of 1912, she wrote her first letter home, exclaiming that she was "fearfully in love" with Debussy and "enraptured" with Schubert (*Ltrs* 1. 66). Before long it became clear that she was also fearfully infatuated with Allen. Sayers named one of her hatpins "Hugh Percy" and a fur stole "Percival," easily associating the name "Percy" with "Percival" since her great-uncle, Percival Leigh, was called "Uncle Percy." For Somerville's 1915 going-down play, the enraptured Dorothy impersonated, with apparent histrionic bravado, Dr. Allen and his conducting style (figures 2.3 and 3.1).[1]

Not only did Sayers perform Hugh Percy Allen with (and on) her body, she also performed him in her writing. In 1914, she composed, hand-printed, and illuminated a thirty-two page poem, "To H. P. A.," as a Christmas gift to Allen. Though Sayers forwarded copies to a few

Figure 3.1 Performing Hugh Percy Allen at Somerville College (Courtesy of The Marion E. Wade Center).

of her Somerville friends, the poem neither reached Allen nor got published—until 1996 when Ralph E. Hone produced an anthology of selected Sayers poems.[2] While "To H. P. A." may have struck Sayers as too gushingly adulatory to send to Allen, she makes oblique allusions to him in poems that were part of her first published book: a collection entitled *OP. I* (1916), which has repeated references to the knight Percival.[3] The poem "A Man Greatly Gifted," which begins "You are the song that a jester sang," is surely addressed to Allen, as is "The Elder Knight." In 1915, Sayers presented the first two sections of "The Elder Knight" to a college friend, titling them instead, "From the Mass in B Minor." Though Hone regards this title as inexplicable, it makes perfect sense if the "elder knight" refers to H. P. A., who led the choir in a performance of the Mass in B Minor which Sayers described as "more exciting than you can possibly imagine" (Ltrs 1. 88).[4]

When Sayers decided to perform, in writing, her excitement over the performances of Hugh Percy Allen, she chose an architectural analogy, comparing Allen to a bell-founder building a tower to house the bell he has newly cast. "To H. P. A." addresses Allen:

> And you that work another prize to win,
> Still fashion for yourself a tower of love,
> With my love set as a small stone therein. (142–44)

The tower seems to represent the Oxford choir, for which Sayers was one voice, or stone, working to sustain the bell-like sounds of Bach's music, which she loved to perform. Just as Sayers had "never seen anything so hot, so tired, so breathless, so excited, or, at the end, so beaming, as Dr Allen" after the "Mass in B Minor" (*Ltrs* 1. 88), so "'Mid throngs of stooping masons moved & burned / The founder's fierce white face & eager eyes" (83–84).

In this poem appears what may be Sayers's earliest celebration of the integrity of work. The tower built "To serve the bell" (104) anticipates a maxim she expressed thirty years later: "the worker's first duty is to *serve the work*" (*Creed* 80).[5] Indeed, when told his tower "is the eighth wonder of the earth," the bell-founder said "He could not build it otherwise" (136, 140). Hence,

> No stone was set within its place unproved,
> He spared not vigil, craft, nor sweat at all
> To build a dwelling for the thing he loved. (86–88)

Sayers has also built a dwelling for the thing she loved: a poem to contain her praise for Hugh Percy Allen, himself a "Master-Builder" (150).[6]

The connection Sayers makes in "To H. P. A." between the work of performance and a work of architecture is not coincidental. From her juvenilia to the last stage of her career, Sayers thought of writing performances as architectural. She even described her intense attraction to *The Divine Comedy* as based upon "the architectural beauty which is the poem's chief glory" (*Further* 36). When she found herself falling in love with Dante in the 1940s, she wrote Charles Williams that "the *Commedia* is like a lodge in a garden of cucumbers." As a "piece of Structure in a sort of jungle of rambling and sprouting vegetation" (*Ltrs* 3. 82), Dante's work, like any performance of genuine creativity, provided for Sayers a container of meaning in the midst of the (cu)cumbersome chaos of life.[7] This perception reflects a powerful doctrine of Sayers's day: that a truly artistic writing performance transforms its theater of representation into a sacred space wherein one can escape the linear temporality of everyday utilitarian language. This chapter will discuss how architectural images perform within Sayers's writings, illuminating not only her own predilections but also the performativity of modernist writings within her imagination.

There is no question but that Sayers sensed some kind of connection between writing and architecture. When analyzing the way people respond to Latin, she metaphorically aligned different perceptions of language with varying architectural styles:

> Everybody is, I suppose, either Classic or Gothic by nature. Either you feel in your bones that buildings should be rectangular boxes with lids to them, or you are moved to the marrow by walls that climb and branch, and break into a [*sic*] inflorescence of pinnacles. (*Search* 184).

When she was working on several writing projects simultaneously, she would place each in a different room of her house, keeping clear their differences in content, it would seem, by associating each with its space of containment. This sensitivity to the poetics of space enabled Sayers to conceptualize three-dimensional settings for her novels, a skill which helped her to envision the architectural and geographical intricacies of Dante's Hell and Purgatory.[8]

In *The Poetics of Space*, Gaston Bachelard discusses how, for many people, books, like architectural spaces, "give our daydreams countless dwelling-spaces," providing the necessary enclosures to "retain our memories" (25, 6). Significantly, when Dorothy L. Sayers attempted to retain her memories in autobiographical form, she spent a considerable amount of time describing the architectural space to which her family moved when she was four years old. A good fifth of "My Edwardian Childhood" focuses upon the features of Bluntisham Rectory (figure 3.2), treating them almost like characters in a story: an "individual and exciting" banister "possessed great personal charm" as it occasionally "exuded a thick, gummy tear of resin," a "knot" sculpted over the mysterious dining room doorway gave the family "food for thought," the kitchen range kept "eating coal." Even more striking than these architectural personifications is the way Sayers identifies the house as text: written in Old English letters on the morning room chimney-piece was "Cheerful Hearts Make Cheerful Hearths" and on the stove in the hall appeared the words "Slow but Sure Combustion" which the young Dorothy loved to "spell out." Even the knot over the dining room doorway looked like a "monogram" that the family "tried to resolve . . . into letters." It would seem that, as Sayers projected her adult sensibilities back into her childhood experiences, she aligned the "beautiful" house with textuality. She even goes so far as to imply that her adult appreciation for Wilkie Collins' *The Moonstone* is related to the architecture of Bluntisham rectory, for she imagines the novel's characters having painted the birds and flowers which adorned the morning room door.[9]

Cat o' Mary duplicates many of these architectural responses and then proceeds to describe the experience of Katherine Lammas at "Beaufort," a fictionalized rendition of Sayers's own years at boarding school. At Beaufort, Katherine discovers words "to dream over" from Malory's *Morte d'Arthur*, comparing them to architecture: "That [quotation] was real, somehow more real than the hall at Beaufort" (101). Feeling that literary art can provide an enclosure for the performance of her imagination, Katherine decides that she will become an author someday, which she renders in architectural terms: "One day I will show them. . . . [and] build, build—something enormous—something they never even dreamed of" (106). Like many writers who reflect upon their own "desire for established existence and a home for consciousness,"[10] Katherine decides she wants to construct imaginative dwelling-spaces for others.

It should come as no surprise, then, that Sayers's earliest known literary efforts contain architectural allusions. When she was seven or eight years old, Dorothy composed (or at least planned to compose) "The Castle by the Sea," of which there is no copy.[11] Her next extant poem, probably written during her early adolescence, makes clear her architectural imagination. Called "The Gargoyle," the sonnet compares a parson who "Spouts" sermons at his congregation with a gargoyle

Figure 3.2 The author in front of Bluntisham Rectory.

that "Spouts at the people in the street" outside. She ends the poem with a pair of witty couplets:

I like the gargoyle best. He plays
So cheerfully on rainy days—
While parsons, no one can deny,
Are awful dampers when they're dry. (11–14)

Employing this architectural image, Dorothy humorously communicates her adolescent ambivalence about things of the church. In fact, up through her first book of poetry, Sayers manifests as much reverential awe for literary art as for religion. To her, as for many of her modernist contemporaries, the texts which stimulated a sense of the sacred were not the dry sermons delivered inside Gothic churches, but the poems which she read inside the Gothic architecture of Oxford University.[12]

Sayers celebrates Oxford in a poem entitled "Lay," first published in *Oxford Poetry 1915*. Like her early adolescent poem, "Lay" expresses a droll affection for gargoyles:

By the clear Cher, thick-clustered dewdrops steep
The heads of mossy gargoyle-beasts, that keep
Their wide, shy smile. (82–84)

Several stanzas later, the gargoyles seem less benevolent—"Odd nightmare carven things, that stare / Spell-stricken in a voiceless scream" (143–44)—perhaps reflecting the "haunted shrines" of a place "so magical / Thou makest me afraid" (108, 49–50). The fear is for the city's loss, coming from one who, having "known thee, worshipped thee, loved thee," wants to hold on to the place and never let go (19). The loss, fulfilled by the narrator's departure from college (124), is imaged through the destruction of its architecture:

Thou makest me afraid,
Lest some great bolt of desolation fall,
And thou in dust be laid
With Babylon and Nineveh the tall;
Or some enchanted lake will cover thee all,
And through quadrangle, cloister, colonnade,
Four-coloured fishes swim, and, faint and small,
Up through the waves at midnight the bells of Magdalen call.
Through midnight waters mighty Tom will call. (50–58)

The bells which call out over the waters echo the bell in Sayers's earlier poem "To H. P. A.": "There, sounding over seas from cape to cape, / The bell rushed to & fro melodiously" (79–80). These bells, protected by their towers, ring in defiance of what Mircea Eliade, in *The Sacred and the Profane*, calls "the formless fluidity of profane space" (63): a water-like fluidity that threatens to engulf sacred space. Indeed, "Water is pre-eminently the slayer; it dissolves, abolishes all form" (Eliade 135).

Sayers's architectural metaphors perform according to the values of her era. Contemporary literary critics distinguished between the formless fluidity of everyday language and the sacred structure of poetry, often developing spatial metaphors to valorize authentic literary art. T. E. Hulme disparaged the "counter language" of everyday diction, calling it an "old pot" that lets meaning "leak out," contrasting it with "the new bowl of metaphor" which contains and retains the "visual" and "concrete." Walter Pater, who whetted the modernist hunger for "concreteness," coined the term "literary architecture" in the late nineteenth century in order to present his belief that "in literary as in all other art, structure is all-important." Following his lead, Cleanth Brooks claimed in 1947 that "the essential structure of a poem (as distinguished from the rational or logical structure of the statement which we abstract from it) resembles that of architecture or painting." Edmund Gosse, whose altar-calls for belletristic art energized university students during the years Sayers attended Oxford, described literature as "this beautiful pinnacled structure of the glory of verse." E. M. Forster used Biblical language to capture poetry's sacred containment: "If there is on earth a house with many mansions, it is the house of words." The architectural metaphor enabled him to distinguish the ephemeral language of "information" from the lasting language of poetry: "Information points to something else. A poem points to nothing but itself. Information is relative. A poem is absolute.... it is eternal and indestructible." As late as 1957, the year Sayers died, famed literary critic Northrop Frye was perpetuating the distinction between sacred literary containment and profane instrumental language:

> In literature, questions of fact or truth are subordinated to the primary literary aim of producing a structure of words for its own sake.... Wherever this autonomous structure is lacking, we have language, words used instrumentally to help human consciousness do or understand something else.

For literary critics during Sayers's lifetime, an artistic poem was an absolute container of human truth, a "verbal icon" toward which they directed their worship.[13]

Basil Blackwell, for whom Sayers worked as an editorial assistant between 1917 and 1919, worshipped accordingly. Written on the endpapers of each book in his "Adventurers All" poetry series, which included *OP. I*, were these words: "It is hoped that these Adventurers may justly claim the attention of those intellects which, in resisting the enervating influence of the novel, look for something of permanent value in the more arduous pursuit of poetry." The dominance of this discourse might explain why Sayers, like her fiction-writing contemporaries C. S. Lewis and Graham Greene, started her literary career publishing poems.[14] All three wanted to be taken seriously as writers, and poetry—not prose—was the sign of a serious writing performance.

Significantly, Sayers begins *OP. I* with an architectural metaphor:

> I will build up my house from the stark foundations,
> If God will give me time enough,
> And search unwearying over the seas and nations
> For stones or better stuff.
>
> Though here be only the mortar and rough-hewn granite,
> I will lay on and not desist
> Till it stand and shine as I dreamed it when I began it,
> Emerald, amethyst.

Sayers here aligns herself with the "Master-Builder" that she created for her tribute "To H. P. A.," who, as we have seen, also works with "stones." Because this poem precedes the volume's Table of Contents, it operates as a comment on the contents themselves, which are mortar and granite making up the "foundation" of a larger "house" of literature that the narrating "I" plans to "build, build" throughout her career. Indeed, the book's title, *OP. I*, signals that this is the first "opus" of many to come, until she constructs an oeuvre that shines with emerald and amethyst. Interestingly, the only poem in *OP. I* with an architectural title, "The Last Castle," contains "tall pillars, dim as amethyst" (33), and one cannot help wondering if Sayers thought of her emerald and amethyst dreams when, nearly two decades later, she placed emeralds in the hammer beams of Fenchurch St. Paul, the architecture she constructs for *The Nine Tailors* (1934).

"The Last Castle" was originally given to two members of The Mutual Admiration Society, a reading group at Somerville College whom the poem explicitly addresses as Sayers laments the ending of their college years:

> O that last faery castle, where we met,
> And dwelt three years together, you and I!
> There was in her no sorrow, no regret
>
> . . . .
>
> Her spires were full of secrets; fairer yet
> Was every bower, and each enchanted room
> When with mysterious rains her courts were wet. (17–19, 22–24)

In these lines, Sayers presents Somerville College as a "protected and protective space" like that described by Victor Brombert in *The Romantic Prison*: a containment which becomes "the locus of reverie and freedom" and hence "of artistic creativity" (5, 15). The creativity generated in the "visionary hall" of Sayers's poem is musical, surely alluding to performances with Hugh Percy Allen: "O marvellously, magically went / Our music up" (37–38). Indeed, in the second stanza of the poem Sayers mentions "Knight Percivale" who gives up his quest for "the Holy Graal" while housed in the "castles" of Oxford. Sayers's sense that the temporality of Percivale's search is brought to rest in spatial containment parallels modernist views of literary art, wherein the temporal search for meaning by "counter language" is brought to spatial fulfillment in the materiality of poetic form.

In 1931, Edmund Wilson, two years younger than Sayers, published *Axel's Castle*, which repeatedly employs architectural images to celebrate writers who, "opposing the flow of Time, establish something outside it: a work of art... an absolute structure." Wilson, like many others, traces this sanctification of poetry back to Coleridge.[15]

Significantly, near the end of "The Last Castle," Sayers alludes to Coleridge's "Kubla Khan," a poem which, according to G. Wilson Knight, contains the ideal sign of "sacred structure."[16] Sayers writes,

> When lo! betwixt the viol and the flute
> One struck a sudden and a haunting sound;
> Each looked upon his neighbour and was mute,
> And let the myrtle branches fall to ground,
> Being afraid; for this was like the noise
> Of something shifting nightly at the root
> Of a tall pleasure-house, whose airy poise
> Rests on the restless quicksand dissolute. (190–97)

This "tall pleasure-house" with "airy poise" reminds us of that "sunny pleasure-dome" to be built "in air" by the narrator of "Kubla Khan." Both pleasure-places elicit memories of haunting music which is now over. In fact, Sayers's castle and Coleridge's dome are similarly described as being on the edge of their existence, bordering on destruction. The "shadow of the dome of pleasure" floated on waves over which Kubla heard "Ancestral voices prophesying war" (29–32),[17] while Sayers's "visionary hall"

> Stood on the limits of our faery land,
> That it and all its treasures, howso rare,
> Must fade at sunrise with the watchman's call. (52–54)

Like that of Coleridge, Sayers's narrator regrets the loss of an idyllic "enchanted" place, and longs to reproduce the enchantment she once felt.[18] Describing a dulcimer-playing Abyssinian maid, Coleridge writes,

> Could I revive within me
> Her symphony and song,
> To such a deep delight 'twould win me,
> That with music loud and long,
> I would build that dome in air
>
> . . . .
>
> And all should cry, Beware! Beware! (41–45, 48)

Coleridge's narrator wants to build a container for the beauty of sound, the pleasure-dome of a poem, cautioning his readers to "Beware" the power generated by such art. Similarly, after reminding her friends of the Xanadu-like "silver streams," "sunshine," and "green fields" surrounding their "last castle," Sayers writes, "beware! for I will let the flood / Of memory burst its banks" (2–6). In response to the potential destruction of architectural enclosure by the exigencies of the temporal,[19] both poets attempt to construct an architectural poem containing their memories. Coleridge builds a "dome in air"—"in air" because poetry is made with the breath of words—and Sayers transforms the "last castle" of her Oxford experience into a poem called "The Last Castle."

Not long after the publication of "The Last Castle," however, Sayers was finding it increasingly difficult to oppose the flow of time. Even though the endpapers for Blackwell's "Adventurers All" series state that "The object of this Series is to remove from the work of young poets the reproach of insolvency," the 26-year-old poet was nearly insolvent. Therefore, in January of 1921 Sayers conceived a plan for a detective tale and in a little over nine months the gestation period was over: the first Peter Wimsey novel, *Whose Body?*, was delivered to the typist by early November. As she baldly put it to one correspondent, "I began writing detective stories in the hope of making enough money to live on" (*Ltrs* 4. 242).[20] In the process, she fell in love with John Cournos, a Russian-born American writer who, like most high modernists, disdained detective fiction as enmired in the temporal. Sayers mentions Cournos in the same letter to her mother that refers to the novel's completion: "John was 'nice' enough . . . , but I fear he has no sympathy with Lord Peter, being the kind of man who takes his writing seriously and spells Art with a capital A" (*Ltrs* 1. 180). Having been published in the cynosure of modernist poetry, Ezra Pound's 1914 anthology *Des Imagistes*, Cournos certainly saw himself engaged in "the more arduous pursuit of poetry" valorized by Blackwell and his modernist peers.[21]

Similar to Cournos, E. M. Forster "openly avowed himself 'too priggish' to enjoy detective stories," while Edmund Wilson spitefully dismissed the genre as "sub-literary," "ill-written," and "degrading to the intelligence."[22] After all, mystery stories were dependent on their linearity, on revelations made through time, rather than on language that drew attention to its own "corporeality." For Sigurd Burckhardt, as for his modernist predecessors, the true literary artist has the ability to "drive a wedge between words and their meanings, [to] lessen as much as possible their designatory force and thereby inhibit our all too ready flight from them to the things they point to."[23] Detective fiction, however, is dependent upon words that point beyond themselves toward clues waiting to be gathered up and laid on the altar of logical analysis.

The modernist derogation of detective fiction bred in Sayers a deep ambivalence about her nonpoetic writing performances. In a 1937 essay that recounts the development of her Lord Peter fictions, "Gaudy Night," Sayers acknowledges that the "detective story of that period enjoyed a pretty poor reputation, and was not expected to contain anything that could be mistaken for 'serious reading.' "[24] It is no wonder that when she develops her literary theory in *The Mind of the Maker*, she describes detective fiction as a simplistic writing performance, employing a metaphor from theater to communicate the artificiality of the genre: "The beautiful finality with which the curtain rings down on the close of the investigation conceals from the reader that no part of the 'problem' has been 'solved' *except that part which was presented in problematic terms.*" The true artist, in contrast, "does not see life as a problem to be solved" (*Mind* 188, 189). Problem solving is a linear, temporal activity, reducing detective fiction to mere craft, far removed from the "absolute structure" of writing performed in theaters of artistic representation.

Sayers therefore came to see herself not as a literary artist but as an artisan. During the composition of her third novel, *Unnatural Death*, she wrote to Cournos about the "craft" of detective fiction:

> It is almost as satisfying as working with one's hands. It is rather like laying a mosaic—putting each piece apparently meaningless and detached—into its place, until one suddenly sees the thing as a consistent picture. (*Ltrs* 1. 241)

By reducing herself to a blue-collar worker in the contemporary hierarchy of literary vocations, Sayers can take pride in her writing and thus justify her pleasure in the performance. And, despite the contempt of her modernist peers, she had already started writing her second detective novel, *Clouds of Witness*, before *Whose Body?* was published. Even when she became a celebrated author, she still thought of herself as performing work with her hands, paralleling the crafting of her plays *The Man Born to Be King* and *The Just Vengeance* to the construction of shelves (*Ltrs* 3. 258).

In "Gaudy Night," Sayers was able to lay out a mosaic that favorably pictured her development as a detective novelist. Having given up on her identity as one who will "build, build" the sacred space of poetry, she nevertheless asserts that she finally produced "a serious 'criticism of life'" in her fifth novel, *The Documents in the Case* (1930), and that her tenth novel, *The Nine Tailors* (1934), was "pretty nearly right," for it combined "detection with poetic romance" (GN 210). By "poetic romance," of course, Sayers does not mean the beauties of falling in love; instead she refers to the "Romance" genre, usually identified with Medieval tales of knights—like Percival. All that is medieval about *The Nine Tailors*, however, is the architecture described within it, architecture that surrounds and reveals the mysteries of the narrative, enabling the novel to approximate the modernist ideals for literary art that Sayers had attempted to fulfill in her poetic writing performances. Consonant, then, with Marvin Carlson's sense of the "operative concept of performance itself," wherein behavior "simultaneously reinscribes and resists pre-existing models," Sayers's employment of a modernist symbol within the antimodernist genre of detective fiction becomes a writing performance that both reinscribes and resists the literary scripts of her day.[25]

Significantly, Peter Wimsey's response to the church nave in *The Nine Tailors* echoes Sayers's poetic description of Oxford's architectural glories. In "The Last Castle" she writes,

> And those tall pillars, dim as amethyst,
> Soaring like smoke incredibly aloof,
> Where, lift on high above the censer-mist,
> Pale capitals glimmered in the golden roof— (33–36)

Almost twenty years later, Sayers repeats the phrase "incredibly aloof" as she describes Peter's "wonder and delight" in Fenchurch St. Paul:

> Then his gaze, returning to the nave, followed the strong yet slender shafting that sprang fountain-like from floor to foliated column-head.... Incredibly aloof, flinging back the light in a dusky shimmer of bright hair and gilded outspread wings, soared the ranked angels.... (34–35)

The one ceiling is "golden," whereas the other is glimmering with "gilded" wings, and both are filled with music. In Fenchurch St. Paul the "shrill voices of the surpliced choir mounted to the roof, and seemed to find their echo in the golden mouths of the angels" (35). Both "The Last Castle" and *The Nine Tailors*, like the architecture they describe, become containers of poetic song, eliciting images of precious gems.

Sayers called *The Nine Tailors*, "pretty nearly right," I would argue, because it demonstrates the architectural containment valorized by modernist literary critics. The church of Fenchurch St. Paul operates as a synecdoche—a part that represents the whole (novel)—projecting the artistry of its spatial form on the temporality of the narrative that contains it. The novel, in fact, begins by contrasting the temporality of a mechanistic device with the spatial form of architecture. The narrative's opening words, "That's torn it!," express Peter's exasperation about crashing his car into a ditch. The machine is "helpless and ridiculous," especially in comparison to the overwhelming solidity of the Medieval church, which elicits from Peter the exclamation "Great Heavens!" five pages later (3, 8).

Despite the fact that she enjoyed riding motorcycles, Sayers shared with her modernist peers suspicions about the temporal benefits of motorized machinery. In 1917, Virginia and Leonard Woolf defied the automated publishing industry by founding Hogarth Press with a single hand-press, reflecting Bloomsbury sensibilities that the sacred realm of belles-lettres is sullied by industrial mechanisms. In 1926, while Hogarth Press was still at its height, Sayers read a book called *Ouroboros: or the Mechanical Extension of Mankind*, by Garet Garrett, which influenced her greatly. In her introduction to *A Time Is Born*, a later work by Garrett (1945), Sayers recounts that Garrett's earlier text "explained why it was that Man, confronted by the machines he had himself invented, stood helpless and bewildered" (v). Garrett's concern about machinery parallels that of I. A. Richards, who wrote three years after *Ouroboros*, "It is arguable that mechanical inventions . . . are disturbing throughout the world the whole order of human mentality, that our minds are, as it were, becoming of an inferior shape."[26] Sayers seems to have agreed, believing that if more humans in industrial society were committed to creatively engaging with fulfilling work, "we could scrap a very great number of the machines, and the 'problem' of industrialization would assume a different aspect" (*Mind* 220). She even staged her own deficiencies as an author in mechanical terms, describing *The Five Red Herrings* as a "machine-made story and no more" (*Ltrs* 4. 303).[27] And in her last detective novel, *Busman's Honeymoon*, Sayers aligns machines with the murderer, having him work in and live over a garage. As Nancy-Lou Patterson notes, he becomes part of an "industrial motif" that starkly contrasts with the "medieval past" of the house in which the Wimseys consummate their marriage.[28]

It is therefore not unique, either for Sayers or her artistic peers, when mechanical exigencies are used to foreground the contrasting stability of an architectural emblem of sacred containment. In *The Nine Tailors*, not only does Peter's car end up in a wreck, but the gates of a sluice fail, and William Thoday is crushed by a barge. The medieval house of God at Fenchurch St. Paul, however, fufills the image of sacred space as described by Mircea Eliade in *The Sacred and the Profane*: the "house is not an object, a 'machine to live in'; it is the universe that man constructs for himself" (56).[29]

Sayers, then, gives form to the temporal mechanisms of her detective plot when she builds a symmetrical structure into *The Nine Tailors*. Peter Wimsey visits Fenchurch St. Paul three times in the course of the novel, and in the middle of his second visit he finds, inside a cope-chest, old bell-ropes. Concentrically situated inside a chest, inside a church, within the architectonic center of the novel, these ropes become clues—central to solving the mystery of a defaced man's death. The word "clue," of course, comes from the ball of thin rope, or "clew," that Ariadne provided for Theseus to help him make his way through the Labyrinth of Minos. Sayers plays upon the etymology of "clue" as Peter reflects on the murdered body during its funeral: "He was amazed that he had not earlier seized upon the rope as a clue to the labyrinth. For the tying-up of the dead man implied so much. Where had the rope come from?" As Peter's reflections proceed, he aligns murder with mechanized transport, like his "car" that had earlier crashed outside Fenchurch St. Paul: "For the body must have been brought from somewhere—how? Car, lorry, cart, wagon, wheelbarrow, truck . . . ? It reminded one of 'Tinker, tailor . . .' " (131, ellipses hers). This last word of Sayers's paragraph—tailor—gives a clue to the reader, for the eight bells, including Tailor Paul, caused the death of Deacon who had been tied up in the belfry during a nine-hour course of bells.

Significantly, the bells, as well as the tower that contains them, are inscribed with text.[30] One can enter the sacred space of the church to read the writing, just as one can enter the novel and seek to interpret the clues that are key to the mystery. The novel is even named after the action of one of the church bells, Tailor Paul, which rings the sign of death, as is noted in writing on its side: "NINE + TAYLERS + MAKE + A + MANNE + IN + CHRIST + IS + DETH + ATT + END + IN + ADAM + YAT + BEGANNE + 1614." This inscription is repeated twice in the novel (76, 118), as though in recognition that death signals—like a bell—the detective novel genre: *Nine Tailors* makes a dead man, whose identity must be solved. But Sayers gives us more than the "final, complete and sole possible solution" of generic detective fiction (*Mind* 188). She creates for her novel the same charged image she had placed in her poems "To H. P. A." and "Lay" almost twenty years earlier: bells, contained by architectural form, which defy "the formless fluidity of profane space."[31] Not only do the bells in *Nine Tailors* kill the criminal, whose actions make him "profane," but they signal the need for sanctuary—sacred space—as they ring out warnings about the fluidity of flood waters advancing on the town. Thus, the church and its bells once again prove to be synecdochic: while the residents of Fenchurch St. Paul take refuge in its sacred space from the profane temporality of the flood, we as readers take refuge from the profane temporality of everyday discourse by entering a well-constructed work of fiction. The novel has fulfilled Walter Pater's definition of literary architecture: "a refuge, a sort of cloistral refuge, from a certain vulgarity in the actual world."[32] Significantly, the last sentence of the novel reads, " 'I'll wish you good-morning, gentlemen,' he said, and went out," signaling an exit from architectural containment at the very moment that the reader exits the text.

Charles Williams was so impressed with the artistry of *The Nine Tailors* that he recommended Sayers as a potential playwright for the annual Canterbury Festival. When the invitation came in autumn 1936, Sayers was taken aback, telling the organizer, Margaret Babington, "it is rather out of my usual line" (*Ltrs* 1. 401). She was a novice at theatrical performances; though she wrote her first staged drama the year before, it was about a subject she knew well—Peter Wimsey—and was done in collaboration with a specialist in drama, Muriel St. Clare Byrne, who taught at the Royal Academy of Dramatic Art. Nevertheless, five weeks later, soon after the collaborative play, *Busman's Honeymoon*, opened to good reviews, Sayers wrote to Babington that she had a plan for the play and two weeks after that they were discussing a possible director. However, Sayers still doubted her abilities, partly because she was intimidated by the writing performances that immediately preceded her. In 1935, Canterbury had produced *Murder in the Cathedral* by T. S. Eliot and in 1936 Charles Williams's *Thomas Cranmer*. Sayers most likely regarded herself as a mere artisan in comparison to Eliot and Williams, who, in her mind, created Art with a "Capital A."

Fortunately for Sayers, artisans were the focus of the 1937 Canterbury celebration, allowing her to write about a little-known figure in comparison to the famous religious leaders in the first two Canterbury performances. Her protagonist, William of Sens, was the architect who designed and coordinated the reconstruction of Canterbury Cathedral after its choir burned down in 1174. Sayers seemed to have identified with William; she certainly regarded writing as architectural, herself as someone compelled to "build, build" rather than to be a religious leader.[33] Significantly, of the members of the cathedral chapter she created for her play, *The Zeal of Thy House*, the one most sympathetic to William is a writer, Father Gervase, who responds to an architectural drawing of William's with "It is wonderful. It is like a poem in stone."[34] The parallel between architecture and literature, builder and writer, is reinforced in a conversation between the two. Gervase despairs, "I wish I could do more to help. But my hands are no good for anything but writing." William responds with a validation of any work that is well done: "If every one would make good use of his own talent and let others do the same, the world would move faster" (39). Gervase later acknowledges that William "thinks of nothing, lives for nothing, but the integrity of his work" (46).

The phrase "integrity of the work" recurs throughout the later stages of Sayers's writing, containing significant polysemy. It can refer to the action of working with integrity, as well as to the integrity of the final product, "the work" of art: performance in both its temporal and spatial meanings.[35] The two, Sayers implies, are interdependent; when one works with integrity she will create a work with integrity. And conversely—as Sayers asserted in her 1942 speech "Why Work?"—"as soon as nothing is demanded of the thing made but its own integral perfection, its own absolute value, the skill and labor of the worker are fully employed and likewise acquire an absolute value" (*Creed* 71–72).

Like her contemporary E. M. Forster, who regarded a poem as "absolute," Sayers seems to reflect a modernist mystification of art. However, rather than substituting the Artist for God as a type of Nietzschean Ubermensch, Sayers, unlike her literary peers, came to see artists as celebrating the nature of God through their work. She actually gives this insight to the Prior in *Zeal*. When Father Theodatus complains about the lack of truth in William, who seduces women and bribes his suppliers, the Prior responds,

> You must not say, without truth,
> Lest you should hear the very stones cry out
> Against you. . . .
> . . . all the truth of the craftsman is in his craft.
> Where there is truth, there is God. (59)

The allusion to stones aligns William with Jesus, who told the Pharisees during his Triumphal Entry in Jerusalem that if those who praised him were silenced, "the stones would immediately cry out" (Luke 19:40). In William's Canterbury Cathedral, the stones cry out about the quality of the work in which they have been placed—reminding us of Sayers's poem "To H. P. A." where her voice performs as one stone in Hugh Percy Allen's "tower of love." The integrity of the worker is proclaimed by the thing made: "when the mouth is dumb, the work shall speak / And save the workman" (*Zeal* 38).

Sayers has the work of Medieval architecture "speak" in both *The Nine Tailors* and *The Zeal of Thy House*, symbolically cleansing the sin contained with its sacred space. In the novel, the bigamist Geoffrey Deacon has hidden stolen emeralds in the beautiful Fenchurch St. Paul roof beams, while in the play the promiscuous William of Sens manifests overweening pride in the beauty of his work. As a result, participants in the life of the church contribute to the punishment of each sinner through the handling of rope. While the rope-pulling bell ringers in the novel unknowingly cause the horrific death of Deacon, who is tied up in the bell tower, the failure of two helpers in the play to inspect a rope suspending William in a "*machine* of [his] own devising" causes his fall (*Zeal* 65, emphasis mine).

The crippled William, loath to give up his work, is redeemed only when he admits that glory, power, and honor do not reside in him, but only in the performance of his work. William prays,

> But let my work, all that was good in me,
> All that was God, stand up and live and grow.
> The work is sound, Lord God, no rottenness there—
> Only in me. Wipe out my name from men
> But not my work . . . (99)

Because most people watching or reading a performance of *Zeal* for the first time were probably unfamiliar with William of Sens, Sayers cleverly suggests that his prayer has been answered: though William's name has been wiped out, Canterbury Cathedral still stands as a famous performance, a "poem in stone."

Not all who enter value the performance, however. In "Why Work?" Sayers tells of someone who, after attending a production of *Zeal*, asked whether Sayers "selected the actors who played the angels for the excellence of their moral character." Sayers, of course, in her commitment to the integrity of the work, believed that "the right kind of actor with no morals would give a far more reverent and seemly performance than a saintly actor with the wrong technical qualifications" (*Creed* 79, 80). She restrained herself from mentioning that the pious old lady echoed one of the least admirable characters in the play, Theodatus, who tells the prior, "I would rather have / A worse-built church with a more virtuous builder." Sayers has the Prior rebuke Theodatus, once again employing an architectural metaphor:

> Will you not let God manage His own business?
> He was a carpenter, and knows His trade
> Better, perhaps, than we do, having had
> Some centuries of experience.[36]

Sayers's fascination with architecture thus affected the way she imagined the performance of God: as both carpenter and "Eternal Architect" (*Zeal* 38). When she has William address Christ as "Architect, poet, maker" (99) and God as the "cunning craftsman" (91), Sayers establishes that William, himself a cunning craftsman, fulfills the *Imago Dei*: performing the image of God. Significantly, over a decade earlier, Sayers had described her own writing performances as the "exercise of cunning craftmanship" (*Ltrs* 1. 241). William, then, is to Sayers as she is to God: a creation who performs as "architect, poet, and maker."

In 1937, the year *The Zeal of Thy House* was first performed, Sayers stated that Peter Wimsey was "a permanent resident in the house of my mind" (GN 220); however, she never published another detective novel.[37] Instead, she began constructing houses of thought for those seeking the Eternal Architect. As one of the wisemen says in her 1938 radio play, *He That Should Come*,

> We build the house of thought, stone upon stone,
> And just as we have finished the topmost pinnacle
> There comes a grinning doubt and pulls away the foundation. (135)

Just as Sayers reached the topmost pinnacle of her detective fiction with the publication of *The Nine Tailors* and *Gaudy Night*, there came a grinning doubt about the worth of the work.

Therefore, after penning several plays on religious material, Sayers started theorizing the worth of artistic construction itself. As a result, she published a work of critical analysis that is still being cited today, sometimes for its theology, other times for its literary criticism: *The Mind of the Maker*, inspired, Sayers admits, by *The Zeal of Thy House*. Anticipating the scripts of performance theorists who assert "that in the postmodern world complicity and subversion are inextricably intertwined,"[38] Sayers will write in *The Mind of the Maker*, "The good that emerges from a conflict of values cannot arise from the total condemnation or destruction of one set of values, but only from the building of a new value, sustained, like an arch, by the tension of the original two" (191). As we will see, *The Mind of the Maker* is sustained, like an arch, by tensions between the modernist sacralizing of art and premodern assurance in a Trinitarian God, thus complying with and subverting the performativity of discourse in her day.

Significantly, performance theorist Marvin Carlson borrows an architectural metaphor from Derrida to comment upon subversive performances :

> Resistant theorists and performers have been very much aware of Derrida's warning that "by repeating what is implicit in the founding concept . . . by using against the edifice the instruments or stones available in the house . . . one risks ceaselessly confirming, consolidating . . . that which one allegedly deconstructs.[39]

Sayers seeks to deconstruct neither literary criticism nor Christian theology. Instead, as we shall see, she uses the founding concept of one discourse to remodel the edifice of another, believing, like Audre Lorde, that "the master's tools will never dismantle the master's house."[40] Assenting to the Gospel writer's assertion that "In the beginning was the word" (John 1:1), Sayers regarded any truly creative performance as a writing performance, one that re-Words the *Imago Dei.*

Chapter 4

Minding the Performance: Sayers's Literary Criticism

I make the wonderful carven beams
Of cedar and of oak
To build King Solomon's house of dreams
With many a hammer-stroke,
And the gilded, wide-winged cherubims.

—Sayers, "The Carpenter's Son"

In her 1918 poem "The Carpenter's Son," Dorothy L. Sayers imagines Jesus contributing to an architectural project. Narrated by a first person "I," the three-stanza poem implies a parallel between the speaking Jesus and the performing poet, between the "house of dreams" and the poetic construction that contains it. Fifteen years later Sayers was indeed building a house of dreams, the gothic church within *The Nine Tailors*, filling it with guilded, wide-winged cherubim. In *The Zeal of Thy House*, angels took on flesh, with the Archangel Michael outlining the threefold work of creation: "an earthly trinity to match the heavenly."[1] In her Canterbury play, Sayers was thus beginning to theorize the creative process itself, an analysis later elaborated through a correspondence with Father Herbert Kelly, who had read a published version of *Zeal* (*Ltrs* 2. 42–50). The result was a "curious" work, as Sayers repeatedly describes *The Mind of the Maker*, in which she presents literary criticism and theology "as two expressions of a single experience" (*Ltrs* 2. 247): two kinds of writing performances staged in the same theater of representation. Significantly, Sayers quotes "The Carpenter's Son" toward the end of *The Mind of the Maker* in order to illustrate a continuity in her own writing performances. She asserts that, just as *The Zeal of Thy House* echoes an important theme in *Gaudy Night, The Mind of the Maker*, as a "hymn to the Master Maker," echoes her "youthful" poem (*Mind* 207). For Sayers, God as "maker" is the master of creative performance, providing a model for all writing performances to follow.

Like the master maker in "The Carpenter's Son" who carves beams as part of a larger project—"King Solomon's house of dreams"—Sayers carved out *The Mind of the Maker* as part of a larger project dreamed up with some friends in 1939: a series of books encouraging "the creative spirit which enables man to *build*" systems of thought and action that meet, even in the midst of war conditions, "spiritual, intellectual and social needs."[2] Though only two other books were completed for the series, Sayers's work, in many people's minds, made the *Bridgeheads* dream worth while. C. S. Lewis called *The Mind of the Maker* "indispensible" while dramatist Maurice Brown compared it to Aristotle's *Poetics.*[3]

Brown's comparison is a bit histrionic. Sayers was not a philosopher by any stretch of the imagination, and *The Mind of the Maker* makes gestures almost as excessive as those Sayers liked to perform on her body. Many readers today would feel uncomfortable with Sayers's opening chapter, which makes assured statements about the unchanging "facts of human nature," which prove that Christian doctrine is "true" (15). However, Sayers's emphatic assertions must be read in the light of their context: written by an Anglo-Catholic during the Nazi blitz. Just as confident demarcations between good and evil became more pronounced in Anglo-American discourse after the bombing of the New York World Trade Center in 2001, Sayers reflected a clear sense that, despite inevitable cultural differences in ethical codes, there is a fundamental "natural law," which states that mass murder is wrong. She grounds this "fact" in lived "experience" (*Mind* 9, 17), paving the way for the rest of her essay, which asserts that the human "experience" of creativity tells us something about the nature of God.

Though Sayers's theory is still being cited by theologians today, the accessibility of her style can make *The Mind of the Maker* seem glib to readers unsympathetic with her Anglo-Catholic assumptions.[4] Sayers's critical insight shines, however, when held up to the light of Mikhail Bakhtin, who also situated his views in "concrete lived experience, its purely aesthetic convincingness."[5] While it is only coincidental that Bakhtin wrote *Author and Hero in Aesthetic Activity* just as Lord Peter Wimsey commenced his heroic activities in the early 1920s, a comparison between the literary theories of these two Christian authors, as distinguished from other critical paradigms of the twentieth century, provides evidence for the sophistication with which Sayers staged her writing performances.

Creation is a key term for both the Anglo-Catholic Sayers and the Eastern Orthodox Bakhtin.[6] Vadim Liapunov, who translated *Author and Hero* from the Russian, explains in his notes that the "basal category for Bakhtin is creation or creative activity," and he cites Russian religious philosopher Nikolai Berdyaev (1874–1948) as a possible influence on Bakhtin's thought (*Author* 234, nt. 16). Significantly, Sayers quotes Berdyaev several times, not only in *The Mind of the Maker*, but also in her essays and correspondence. She explicitly attributes her sense of the *Imago Dei*—the image of God manifest in humans—to Beryaev's *The Destiny of Man*, which provides an epigraph for the fifth chapter of *Mind*:

> God created man in his own image and likeness, i.e. made him a creator too, calling him to free spontaneous activity and not to formal obedience to His power. Free creativeness is the creature's answer to the great call of its creator. (61)

For Sayers, humans fulfill their created destiny most completely through their own acts of creation. Bakhtin regards this destiny as "the *divinity of the artist*": "the artist and art as a whole create a completely new vision of the world, a new image of the world, a new reality of the world's mortal flesh" (*Author* 191).

The Incarnation directs the aesthetic theories of both Bakhtin and Sayers, who regard literary creation in trinitarian terms. For Sayers, the material form of a work of art, like the body of Jesus, is the "Energy" or "Activity" that proceeds from the "Idea" of the Creator-Author, generating "Power," as does the Holy Spirit, through the response of the beholder-reader. At the simplest level, "Idea" corresponds to a Book-as-Thought, "Activity" to a Book-as-Written, "Power" to the Book-as-Read (*Mind* 122). However, it would do disservice to the complexity of Sayers's thought to limit her trinitarian aesthetic to such bald terms, for elsewhere in *The Mind of the Maker* she establishes that Idea, Energy, and Power, like Father, Son, and Holy Ghost, are simultaneously engaged and mutually interdependent rather than temporally linear and hierarchically structured:

> The Idea, that is, cannot be said to precede the Energy in time, because (so far as that act of creation is concerned) it is the Energy that creates

> the time-process. . . . The writer cannot even be conscious of his Idea except by the working of the Energy which formulates it to himself. (*Mind* 38–39)

Bakhtin makes a very similar point: "An author creates, but he sees his own creating only in the object to which he is giving form, that is, he sees only the emerging product of creation and not the inner, psychologically determinate, process of creation" (*Author* 6). In other words, the author cannot separate her creative Idea from the incarnated form, or Hero, which becomes manifest through the performance of writing.

To regard thought as preceding language—the idea as preceding its writing performance—is to reflect an Arian view of creation, wherein God created the Son. For both Bakhtin and Sayers, the Hero/Son is begotten, not made. This dogma of Christian orthodoxy—formulated at the Council of Nicaea in 325 A.D. and performed in Sayers's last written play, *The Emperor Constantine*—may be what enabled both theorists to formulate an aesthetic in defiance of modernist discourse. As David Haney explains in his book on the hermeneutics of incarnation,

> [T]he doctrine of incarnation does *not* presume the distinction between body and spirit that is retranslated in both Enlightenment and modern thought into a conventional relationship between two isolable units that can be labeled *sign* and *idea* (Locke) or *signifier* and *signified* (Saussure).[7]

For Bakhtin and Sayers, writing does not give flesh to an antecedent idea; instead, just as Jesus's performances on earth express, simultaneously, the performativity of God, a writing performance simultaneously performs thought. As a result of their premodern assumptions, then, the literary theories of Sayers and Bakhtin anticipate postmodern views of textuality, wherein thought cannot be divorced from the language that embeds it.

With the incarnation as the basis of their aesthetic, Sayers and Bakhtin regard writing performances in humanizing, rather than objectifying, terms. Bakhtin states, "spatial form is not *sensu stricto* the form of a work as an object, but the form of a hero and his world" which is in "relationship" with the Author-Creator (*Author* 89). Bakhtin's aesthetic as well as his ethics are predicated upon the notion of "outsideness": the "other" can be consummated, or brought to completion, only by a self/author that stands outside of the other/hero's position, where he or she can see what the other cannot see of their own body, environment, and/or life circumstances. The author, then,

> uses all the privileges of his all-round position outside the hero. At the same time, the hero in this form of interrelationship is, in his turn, most independent, alive, aware, and persistent in the cognitive and ethical (value-oriented) position he maintains in living his own life. (*Author* 174)

While, for Bakhtin, the "hero" refers to the product of any writing performance, as does the "Activity" in Sayers's triad, the actual hero of Sayers's detective fiction might nevertheless—if somewhat whimsically—illustrate Bakhtin's paradigm.

In her earliest letters that allude to Lord Peter Wimsey, Sayers's hero seems to be "living his own life," as Bakhtin puts it: "The typist is dawdling over him, it seems to me, but as I haven't the cash handy I bear up! But when he *does* arrive I ought to trot him round to various publishers before giving up altogether" (*Ltrs* 1. 181). When she writes in 1936 "How I Came to Invent the Character of Lord Peter," Sayers refers to him as an autonomous "hero" rather than a literary invention: "My impression is that I was thinking about writing a detective story, and that he walked in, complete with spats, and applied in an airy don't-care-if-I-get-it way for the job of hero."[8] She thus mirrors Bakhtin's sense that "It is this extra-aesthetic reality of the hero that will enter as a shaped reality into the work produced" (*Author* 199). Writing performances generate actors who can walk on and off the stage of representation.

Because Sayers believed in the independent "activity" of an author's creatures, she disdained readers who identified her tastes with those of her hero-creation: "what is intolerable is that the created being should be thus violently stripped of its own precious personality" (*Mind* 130). She also criticized readers of the other extreme: so convinced by the independence of Peter from his author's imagination, that they want him to perform according to their own scripts. Sayers quotes someone saying of Peter, "But he's *far* too intelligent and far too nice, not to be a Christian"; she responds quite peremptorily with the statement "He exists in his own right and not to please you" (*Mind* 131). Sayers has given a tangible illustration of Bakhtin's sense that "[t]he hero's inner directedness from within his own lived life possesses its own immanent necessity, its own autonomy; as such, it is capable of compelling us at times to become involved in its own sphere, in its own becoming" (*Author* 84). In contradistinction to the stasis of "being," Bakhtin valorizes "becoming" as the change generated when a self authentically interacts with the "other," who, in a position of "outsideness," can see things the self overlooks.

Sayers became involved in Peter's "becoming" when she decided to give him a significant "other." Originally, as she admits in her "Gaudy Night" essay, Sayers wrote *Strong Poison* with the intention of "marrying [Peter] off and getting rid of him" (210). However, as soon as she formed Harriet Vane from the rib of Peter's character, Sayers realized that Wimsey needed to mature in order to be worthy of Harriet's love. Hence, when she gave Energy to her Idea about Peter, a new Power was generated: "the puppets had somehow got just so much flesh and blood in them that I could not force them to accept [superficial love] without shocking myself" (GN 211). Sayers experienced how the "activity" of writing can change an author's "idea."

The love that Sayers created for Peter, who refuses to subdue Harriet to his desires, mirrors the author's love for her creation:

> [T]he creator's love for his work is not a greedy possessiveness; he never desires to subdue his work to himself but always to subdue himself to his work. The more genuinely creative he is, the more he will want his work to develop in accordance with its own nature, and to stand independent of himself. (*Mind* 130)

Bakhtin, as well, sees the hero's independent consciousness as lovingly produced by the consciousness of the author:

> In the aesthetic event, we have to do with a *meeting* of two consciousnesses which are in principle distinct from each other;...and it is this, the hero's own consciousness, that is concretely localized and *embodied*...and lovingly *consummated*. (*Author* 89)

As a result of this love, the author, says Sayers, "does not desire that the creature's identity should be merged in his own, nor that his miraculous power should be invoked to wrest the creature from its proper nature" (*Mind* 132). Liapunov's translation of Bakhtin employs the same word "merge" as a warning against imbalanced authorial activity: "Where the author merges with the hero, the form we get is, indeed, no more than pure expression in the sense of 'expressive' aesthetics, i.e., it is the result of the self-activity of the hero in relation to whom we failed to find an exterior position" (*Author* 84)—a failure of the "outsideness" which, according to

Bakhtin, should be modeled on the loving mind of the Maker God: "What I must be for the other, God is for me" (*Author* 56).

Both Sayers and Bakhtin therefore regard the author's relation to the hero as echoing the theological paradox of free will and determinism. For Sayers, the author

> is conscious of the same paradoxical need—namely, the complete independence of the creature, combined with its willing co-operation in his purpose in conformity with the law of its nature. In this insistent need he sees the image of the perfect relation of Creator and creature, and the perfect reconciliation of divine predestination with free created will. (*Mind* 138)

Bakhtin makes a very similar statement in *Problems of Dostoevsky's Poetics*:

> [T]his independence and freedom of a character is precisely what is incorporated into the author's design. This design, as it were, predestines the character for freedom (a relative freedom, of course), and incorporates him as such into the strict and carefully calculated plan of the whole. (13)

For both, God's performance relative to creation provides a model for any author's writing performances.

In *The Mind of the Maker* and *Author and Hero in Aesthetic Activity*, Sayers and Bakhtin spend much less time discussing the third member of their aesthetic trinities, and both have been criticized for it. Ruth Coates takes Bakhtin to task for his "imbalance of power," wherein the "Author" generates a "benign dictatorship based on the reassuring model of the Creator's relationship to his Creation." Nicholas Wolterstorff criticizes Sayers's "Romantic" sense of "the artist alone with images, trying to make them cohere into aesthetic unities." A preferable model of artistic creation, suggests Wolterstorff, is one in which "the work having been made, we all begin to interact with it and it contributes to our flourishing."[9]

The hesitancy of Sayers and Bakhtin to emphasize the role of the reader may reflect the intense animus against considerations of audience among their contemporaries. The critical elite of their time seemed to consider the quality of High Art as inversely proportional to its mass appeal. In painting as well as music and literature, a great work of art did not pander to the average person's desire for representation or reference; instead, it self-reflexively called attention to its own medium—whether through color, tone, or word. In 1939, the same year Sayers was beginning to conceptualize *The Mind of the*

Maker, Clement Greenberg, the famous definer and defender of abstract expressionism, presented his uniquely modernist take on the *Imago Dei*: the "avant-garde poet or artist tries in effect to imitate God by creating something valid solely on its own terms, in the way nature itself is . . . valid: something *given*, increate, independent of meanings, similars, or originals."[10] In contrast, both Sayers and Bakhtin regarded the "contemplator" as key to the power of the work: "to feel the form, to feel its saving power, its axiological weight—to feel its beauty" (*Author* 200). Hence, rather than criticize them for not emphasizing the reader's role more emphatically, perhaps we should be impressed that they recognized its importance at all.

By acknowledging the role of the performing reader, Bakhtin and Sayers were defying, respectively, Russian Formalism and Anglo-American New Criticism: academic literary theories that reinscribed the modernist mystification of autonomy, not so much of the artistic genius (although theorists certainly endorsed such a notion) but of the art object itself. Bakhtin's criticism, scholars agree, was to a certain extent a response to the Russian Formalism that was at its height when he began his writing career. In "The Problem of Content, Material, and Form in Creative Literature" (1924) Bakhtin, as Ruth Coates notes, "criticizes Formalist poetics for reducing literary form to an aspect of the verbal material, and for substituting technical craftsmanship for true aesthetic activity."[11] Bakhtin regarded writing performances as actions to be engaged rather than objects to be observed.

Sayers also eschewed the objectivist commodification of a work of art. For her, the noun "work" alludes to the activity indicated by its verb form—as incarnated by the performances of both author and reader. Although it is highly unlikely she was aware of John Crowe Ransom's book *The New Criticism* while she was writing *The Mind of the Maker*, since both were published the same year (1941), Sayers later describes the New Criticism in a talk delivered to the Oxford University Spectator Club (1956): "There is a school of criticism which asserts roundly that art and communication have nothing to do with one another. A work of art can only be assessed by that which it is in itself, and not by its effect upon the audience or spectators" (*Search* 14). She argues against what New Critics W. K. Wimsatt, Jr., and Monroe Beardsley were to call "the affective fallacy," believing that poetry should, indeed, affect the reader.[12]

Ransom defined the New Criticism partially through and partially against influential literary critic I. A. Richards. Exactly Sayers's age, Richards sounds like a New Critic when he says in *Science and Poetry*

(1926) that "It is never what a poem *says* which matters, but what it *is*" (34–35). Her main problem with *Science and Poetry*, however, was with "what Mr. I. A. Richards has called 'pseudo-statements' " (*Search* 17). Influenced by the logical positivism of the 1920s, Richards held that only empirically verifiable statements can be designated "true." Thus religion, with its primitive "Magical View of the world" (73), must be surmounted:

> Countless pseudo-statements—about God, about the universe, about human nature . . . about the soul, its rank and destiny—have suddenly become, for sincere, honest and informal minds, impossible. For centuries they have been believed; now they are gone, irrecoverably. (71–72)

Science, however, cannot address "ultimate" concerns (63). Richards therefore substitutes the "pseudo-statements" of poetry for those of religion: "Briefly, if we can contrive to believe poetry, then the world seems, while we do so, to be transfigured" (73). Like many critics of his generation, Richards believed that poetry, though formulated of pseudo-statements, "is a perfectly possible means of overcoming chaos" (95). To the woman who wrote "Creed or Chaos," however, Richards's answer to the chaos of the world was naive—much as it is regarded by theorists today.

In her own day, Sayers agreed with R. G. Collingwood's critique of the artificial division between scientific and poetic language advanced by Richards. In *Principles of Art* (1937), Collingwood writes, "Dr. Richards assumes, apparently without realizing that any one could do otherwise, that language is not an activity, but something which is 'used', . . . like a chisel that is used either for cutting wood or for lifting tacks" (262). Sayers likewise affirms language as activity: "Every word is an *event*," not an immutable thing like a chisel: a "word or a phrase is not, and cannot be, an instrument of precision" (*Search* 272, 271). In fact, in her essay "Creative Mind," Sayers aligns language with a rather humorous and quite unscientific activity:

> The desperate attempts of scientists to reduce language to a kind of algebraic formula in which the same symbol has always the same meaning resemble the process of trying to force a large and obstreperous cat into a small basket. As fast as you tuck in the head, the tail comes out; when you have at length confined the hind legs, the forepaws come out and scratch; and when, after a painful struggle, you shut down the lid, the dismal wailings of the imprisoned animal suggest that some

> essential dignity in the creature has been violated and a wrong done to its nature. (*Whimsical* 93)

Anticipating poststructuralists to follow, Sayers recognizes that language, like a cat, directs the performances of those who attempt to control it for their purposes. Therefore, anything created out of language will also, like the cat, have a mind of its own. This, of course, ties into Sayers's trinitarian aesthetic: just as the fully human Jesus, not being a mere "tool" of Creator God, had a mind of his own, so the "Activity" of a literary work, as expressed in language, has a mind of its own. Once again, we see Sayers's theory harmonizing with that of Bakhtin, who regarded "the work of art as a living artistic event . . . and not as something that has been . . . reduced to the bare empirical givenness of a verbal whole" (*Author* 189).

While it is nearly impossible for Sayers to have read Bakhtin, it is almost certain that Collingwood's *Principles of Art* influenced *The Mind of the Maker*.[13] Though Sayers started to formulate her aesthetic theory in 1937, the year *Principles of Art* was first published, it is probable that, as she expanded her idea into a book-length study, she consulted Collingwood's work. Furthermore, although it was Father Kelly who seems to have first pointed out the relevance of the Athanasian Creed to Sayers's trinitarian view of creativity (Ltrs 2. 42 ff.), Collingwood aligns the very same creed with the aesthetics of "creation." Wolterstorff actually criticizes *The Mind of the Maker* for being *too* "Collingwoodian," asserting that Sayers, like Collingwood, believes creation can occur "without any material manifestation." Quite legitimately, he is troubled that Sayers does not say enough about the "demands of audiences and patrons" or "the social realities of art—nor indeed, about its cultural and material realities."[14] Sayers, however, is not silent on these issues, stating that "between the poet and his age there is an intimate connection of mutual influence." Sayers, in fact, endorses "creative citizenship," believing that artists can shock politicians out of their complacency: "At the irruption of the artist into a State department, officialdom stands aghast, not relishing the ruthless realism which goes directly to essentials." A writing performance, believed Sayers, has axiological weight through its effect on the reader, generating "a communication of Power to the world."[15]

Acknowledging the importance of audience, Sayers and Bakhtin foreshadow *Rezeptionsästhetik*, or Reception Theory, identified at its beginnings with the writings of Hans Robert Jauss and Wolfgang Iser. Called a "novel development" in literary history, Reception Theory arose in the 1970s as an answer to Russian Formalism and the New Criticism.[16] Significantly, not only do Iser and Jauss refer to constructs developed by Collingwood as they discuss the reader's role in the work of art,[17] but they also establish trinitarian views of the aesthetic object: the interdependence of origin-work-reader. Iser even implies an incarnational nature to the art work when he endorses J. M. Lotman's view that "The literary text acts like a sort of living organism, which is linked to the reader, and also instructs him, by means of a feedback system" (Iser 66–67). And Jauss sees such instruction as affecting social behavior: "The relationship between literature and reader can actualize itself in the sensorial realm as an incitement to aesthetic perception as well as in the ethical realm as a summons to moral reflection" (Jaus 41). For Sayers, of course, the reader's moral reflection becomes ethical "when it incarnates itself in actions" (*Mind* 111).

Comparable to Sayers's Idea, Activity, and Power, the trinity informing Iser's aesthetic object is repertoire, strategies, and realization. "Realization" refers to the "reader's participation" in the text, bringing it to fulfillment. The word "strategies," designating the "accepted procedures" for inscribing a creative text, gestures in its dynamism toward Bakhtin's "Hero" and Sayers's "Activity." In contrast, "repertoire" is the least like Sayers's and Bakhtin's trinitarian aesthetic. Instead of an Author-Creator, Iser situates as "origin" the conventions that precede but also inhere within the art object (69). Nevertheless, as Susan Suleiman notes in her introduction to *The Reader in the Text*, Iser seems to suggest that, because "the reader's activity of filling in the gaps is 'programmed' by the text itself, . . . the kind of pattern the reader creates for the text is foreseen and intended by the author" (25). Sayers also recognizes the importance of "repertoire" to the authorial "Idea," outlining in *The Mind of the Maker* the repertoire of literary and social norms that inform *The Nine Tailors* (120–22). Illustrating what Wolterstorff calls the "cultural and material realities . . . embodied within the work" (465), Sayers thus suggests that the authorial Idea is always culturally situated, its Activity energized by the Power of other writing performances: a mutual communion of three elements into one new writing performance.

The word "communion" repeatedly occurs in the writings of a theorist who influenced both Iser and Jauss: Roman Ingarden. In his essay on "Phenomenological Aesthetics," Ingarden asserts that, "between the observer and the artist, the master, there arises a specific *reapproachment*, even a certain kind of spiritual communion" (268). And he recognizes that sometimes the observer and the artist are one and the same person: "the artist becomes an observer of his own emerging work, but even then it is not completely passive apprehension but an active, receptive behavior" (261). Sayers sees such "active, receptive behavior" in the artist as resulting from the creative Power of the work: "It is the thing which flows back to the writer from his own activity and makes him, as it were, the reader of his own book" (*Mind* 40–41). For her, the Power enters into and energizes the receiver, becoming part of the creativity of the whole. Ingarden would agree: "[T]he observer too does not behave in a completely passive or receptive way, but being temporarily disposed to the reception and recreation of the work itself, is also not only active, but in a certain sense at least creative" (261). Ingarden describes the recreation enacted by the beholder-receiver as "concretizing" the work, suggesting that, just as a "certain kind of spiritual communion" is achieved "between the observer and the artist," the encounter between the observer and the work might be called "communion" as well (268, 262). He even goes so far as to anthropomorphize the artistic work, as Sayers does when she compares it to Christ the Son and Bakhtin does when he designates it "Hero": "Works of art have *a right to expect* to be properly apprehended by observers who are in communion with them" (269, emphasis mine). With his reiteration that "communion" is necessary to the fulfillment of an art object, Ingarden comes very close to the trinitarian aesthetics of Sayers and Bakhtin.[18]

The point here is not that Sayers or Bakhtin influenced *Rezeptionsästhetik*. It is highly unlikely that Roman Ingarden, Hans Robert Jauss, or Wolfgang Iser read either of the earlier writers. What we see, however, is that the trinitarianism which informs the theories of both Sayers and Bakhtin led them to conclusions similar to those of innovative scholars in the last half of the twentieth century. However, while the Reception Theorists focus primarily upon the reader's role in actualizing the text, "concretizing" it (Ingarden) by filling up its "blanks" (Iser), Sayers and Bakhtin place the emphasis upon the God/Christ relationship of Author and Hero. Whereas Jauss discusses the "horizon of expectations" in the consciousness of the reader, Bakhtin focuses on the Author's "horizon" of "act-performing (forward-looking) consciousness" (*Author* 134).

Postmodern scholars often consider Bakhtin's own act-performing (forward-looking) consciousness to be most fully revealed in his theory of heteroglossia, which challenged the modernist sanctification of artistic autonomy by contextualizing writing performances.[19] For Bakhtin, all utterances are conditioned by the place and time of their performance, uniquely combining the diverse forces—or tongues—within discourse at any particular moment. Bakhtin therefore presents the need for "dialogism": an openness to and interaction among the various meanings generated by heteroglossia—whether manifest in a text or among human beings. Having been created by a Master Maker, humans are inscribed, like works of art, by the numerous tongues of their lived environment—the political, the religious, the ethnic, the sexual, and so on—some of which contradict each other, all of which take on different meanings in different contexts. The heteroglossia of the writer, then, becomes performed in her writing. However, only the best writers successfully dialogize the heteroglossia that contextualizes their work, maintaining a pluralism of perspectives without endorsing only one as "true."

Bakhtin identifies as masters of dialogism two Christian authors: Dante and Dostoevsky. Though he devotes much more time and attention to his fellow Russian writer, he praises Dante for manifesting "an acute feeling for [his] epoch's contradictions.... From this springs that striving toward as full as possible an exposition of all the contradictory multiplicity of the epoch" in his work.[20] Sayers expresses her love for Dante in similar, albeit more appetizing, terms:

> The Middle Ages were not a pigeon-holing period like ours—they mixed their intellect, feelings, philosophy, romance, politics, science and religious beliefs, oratory, small-talk, and contemporary gossip into one magnificent and variegated Christmas pudding and took it down, using whatever spoon came handy, suet and plums together, with an omnivorous appetite at which we can only marvel. The thing that staggers one is that, when one compares him with his contemporaries and even with his successors, Dante should have been able to combine all that vast array of ingredients into anything so simultaneously light, rich, and palatable. (*Ltrs* 3. 498)

Not all readers, of course, enjoy such variegated tastes mixed together in one dish. Instead, they attempt to resolve the conflicts and ambiguities of heteroglossia through what Bakhtin calls "monologism," such that one discourse subsumes all the others, rendering dialogism inert.

Monologism might be illustrated through the genre of detective fiction, the "distinguishing feature" of which, as George Dove notes, is "its resolution, the closure of a mystery that is both compelling and artificial."[21] In *The Mind of the Maker*, Sayers similarly asserts that the "problem and solution" methodology of mystery fiction leads to a "predictable, final, complete and sole possible solution" that readers "often welcome" (187–88). She goes on to show how the scenarios set up by detective stories "falsify our apprehension of life as disastrously as they falsify our apprehension of art" (194). As a result, the early Peter Wimsey was as one-dimensional as the genre that enfolded him—a mere puppet of the author, as Sayers admits in "Gaudy Night":

> I plugged confidently on, putting my puppet through all his tricks and exhibiting him in a number of elegant attitudes. But I had not properly realized—and this shows how far I was from understanding what it was I was trying to do with the detective novel—that any character that remains static except for a repertory of tricks and attitudes is bound to become a monstrous weariness to his maker. (210)

When she was to write *The Mind of the Maker* several years later, Sayers recognized that a hero should be a puppet no more than a human should be. Instead, both provide "a kind of illumination upon the variety and inconclusiveness of the world about us" (*Mind* 52). Having come to desire inconclusiveness even for Lord Peter, Sayers decided to "perform a major operation," turning him into a "complete human being" (GN 211). In other words, she decided to inscribe him with the inconclusiveness of heteroglossia. She thus worked toward the dialogic relationship between author and hero celebrated by Bakhtin, "one that affirms the independence, internal freedom, unfinalizability, and indeterminacy of the hero."[22]

Sayers's shift toward heteroglossia came in 1928 when she set aside her one-dimensional hero in order to write her fifth novel. The absence of Peter may have contributed to Sayers's sense that *The Documents in the Case* was a "jump forward" in her literary abilities, enabling her to create "a serious 'criticism of life'" through the plural perspectives of her modified epistolary format.[23] Though it would be ludicrous to suggest that she got near to approaching the dialogism of Dante and Dostoevsky—the genre of detective fiction made it nearly impossible—we nevertheless can see Sayers struggling to express what Bakhtin calls "a Galilean perception": one that "denies the absolutism of a single and unitary language" (*Dialogic* 366).

In *Documents*, Sayers creates several characters more complex than those in her earlier novels, their strengths and weaknesses revealed not only through the attitudes and perceptions of their own letters, but also in the comments made about them in letters and statements written by others. Through *The Documents in the Case* we see illustrated Bakhtin's sense that, "In the mouth of another person, a word or a definition identical in content would take on another meaning and tone, and would no longer be the truth."[24]

One word taking on different meanings and tones in *Documents* is "love." Margaret Harrison believes in the truth of her love for the painter Lathom, while her husband defines love in a different way. Her statement is true from the position where she is situated—married to an unimaginative, truculent bore—a perception endorsed by Miss Milsom. To other observers, however, her definition of "love" looks like a self-blinding rationalization for adultery.

Heteroglossia becomes dialogized when Paul, Harrison's adult son by his first wife, speaks to Margaret after his father's death. She says of her husband, "I think he had an unhappy nature," only to have Paul dismiss her perception. He reports their conversation in such a way that we feel sympathy for both positions:

> "That is nonsense," I said, "and you know it. My father was a most simple, friendly, companionable man—only you never would be a real wife to him."
>
> "He wouldn't let me," she said.... "He dried me up. He broke my spirit, Paul."
>
> "My father was not a demonstrative man," I said, "but you know quite well that he was proud of you and devoted to you. If you had heard him speak of you as I have..."
>
> "Ah!" she said, quickly, "but I never did. That was the trouble. What is the good of being praised behind one's back if one is always being scolded and snubbed to one's face?" (152)

Sayers makes clear from the documents of the case that both Margaret and Paul are right in their conflicting assessments of Mr. Harrison. The condition of growing up as Harrison's son, and becoming an engineer like him, makes Paul's utterances about companionability and devotion mean something radically different than those made by his father's much younger, emotionally hungry wife. Sayers thus illustrates the dialogic imagination described by Bakhtin:

> [T]he author utilizes now one language, now another, in order to avoid giving himself up wholly to either of them; he makes use of this verbal give-and-take, this dialogue of languages at every point in his work, in order that he himself might remain as it were neutral with regard to language. (*Dialogic* 314).

Sayers extended her neutrality in *The Documents in the Case* even to her criminal, making Lathom, her adulterous murderer, more engaging than either the murdered Harrison or his son. While in most Golden Age detective fiction, the amateur investigator elicits the reader's interest, if not admiration, Paul Harrison generates very little sympathy due to his contemptuous, monologic dismissal of almost every idea and every person that he encounters. In contrast, as Sayers tells Eustace Barton, her collaborator, "The murderer...has my sympathy" (*Ltrs* 1. 302). Sayers thus shows that seeing is always situated, even the seeing of guilt and innocence.

This is not to imply that Sayers thinks Lathom should be exonerated for his murder; she simply recognizes that he is more than his act of murder, that he is "unfinalizable." His real sin, in Bakhtinian terms, is refusing to allow unfinalizability for Harrison, literalized when he reduces Harrison to the finality of death. However, Sayers implies that Margaret Harrison is equally culpable. Though not committing the actual murder, Margaret has finalized her husband by refusing to see his positive qualities. Sayers has Margaret refer to Harrison as "the Gorgon" with his "cold stony eye" (98). In classical tradition, the Gorgons, known to literally petrify their adversaries, were female; therefore, if Harrison is as cold as stone, the true Gorgon here, as Sayers hints, is Margaret, who has finalized her husband as "selfish" and a "tyrant" (101, 107). And when she manipulates Lathom to be her "Perseus" and kill the "stony horror" of her husband (99), Margaret also finalizes Lathom, who dies for his crime. Not coincidentally, Sayers has Margaret repeatedly call Lathom "Petra," a play on the Greek word for stone.

Because so few people in *The Documents in the Case* are charitable toward each other, the novel, as Sayers despaired upon its completion, "is turning out rather grim and sordid" with "an atmosphere of tense depression" (*Ltrs* 1. 299). She even has Munting admit to his fiancée how difficult it is to encourage the unfinalizability of others. He worries that, once they are married, "I shall try to mould you into the mirror of myself, fatally and inevitably" (35). At least Munting recognizes his biased self-interest. The other writers of *The Documents in the Case* remain blind to their biases, as when Agatha Milsom quite myopically writes, "I (not having my eyes blinded) can see quite clearly" (47). Significantly, when Munting asks a group of scientists, "What is Life?" a chemist replies that "it is a kind of bias—a lop-sidedness" (214).

Because every life is situated in a particular context, it can develop in lopsided ways, seeing according to individual bias. As Bakhtin notes,

> When I contemplate a whole human being who is situated outside and over against me, our concrete, actually experienced horizons do not

> coincide. For at each given moment, regardless of the position and the proximity to me of this other human being whom I am contemplating, I shall always see and know something that he, from his place outside and over against me, cannot see himself: parts of his body that are inaccessible to his own gaze (his head, his face and its expression).... As we gaze at each other, two different worlds are reflected in the pupils of our eyes. (*Author* 22–3)

In *The Mind of the Maker*, Sayers similarly focuses on eyes in order to discuss the lopsidedness of perception:

> [W]e cannot follow the movement of our own eyes in a mirror. We can, by turning our head, observe them in this position and in that position with respect to our body, but never in the act of moving themselves from one position to the other, and never in the act of gazing at anything but the mirror. Thus our idea of ourself is bound to be falsified, since what to others appears the most lively and mobile part of ourself, appears to us unnaturally fixed. (115)

For Bakhtin, the lopsidedness of life is ameliorated through the "excess of seeing" generated when humans consummate, rather than finalize, each other:

> I must empathize or project myself into this other human being, see his world axiologically from within him as *he* sees this world; I must put myself in his place and then, after returning to my own place, "fill in" his horizon through that excess of seeing which opens out from this, my own, place outside him (*Author* 25).

In *The Documents in the Case*, Sayers asserts the need for such consummation through her one admirable character, Rev. Perry, who states, "We suffer for one another, as, indeed, we must, being all members one of another" (211). When a biologist dismisses this comment, saying Perry is trying "to keep up that story about Man in the image of God," we see Sayers gesturing, a decade in advance, toward the construct foundational to *The Mind of the Maker*: the image of God manifest in human creativity.

In fact, the creation of life becomes a key motif in *The Documents in the Case*. Lathom, while painting life portraits of Margaret Harrison and Agatha Milsom, creates new life in Margaret's womb. Munting, while writing the "life" of a Victorian author, thinks about making a new life in marriage: "I must finish the *Life* and then get on to *our* life" (24). Meanwhile, he struggles over scientific and religious explanations for the origins of life, eliciting from Paul, "You're cracked about the origin of life. It's the origin of death we're investigating" (205). The threads of Sayers's story come together at a dinner party hosted by Perry, at which time Munting, in conversation with several scientists, is presented with clues to the death of Harrison as well as clues to the creation of life.

Sayers names the scientist most responsible for solving the mystery of Harrison's death "Waters," suggesting that in "waters" we may also get at the mystery of life: " 'Well,' said Waters, '[the author of Genesis] put the beginnings of life on the face of the waters, which wasn't so very far off' " (212). He goes on to affirm a concept of intelligent design behind the universe: "at the moment when Life first arose, a directive force came into play"—a first cause that, as another scientist puts it, "produced the first asymmetric molecules of Life" (216, 217). Significantly, a search for the asymmetric molecules of Life determines the cause of Harrison's death. Waters coordinates an experiment to assess whether Harrison ingested the asymmetric compound of poisonous mushrooms he unintentionally gathered, or whether Lathom intentionally poisoned him with the symmetric compound of inorganic synthetic muscarine. Immediately before the experiment is begun, Sayers has Munting hum to himself bars from Haydn's *Creation*, complete with the drum sounds: "And-the-spi-rit-of-God (pomty) moved-upon-the-face-of-the-waters (pom)" (220).

Upon the face of Waters during the experiment, however, Munting sees an image of the dead Harrison:

> [Waters] snapped off the lights, and we were left with only the sodium flame. In that green, sick glare a face floated close to mine—a corpse-face.... I pulled myself together and realised that I must look just as ghastly to him as he to me. But for the moment the face had been Harrison's... (225)

Up until this point, Munting had resisted believing Lantham was the murderer, wanting to think the best of the brilliant, affable artist, who was so much easier to like than Harrison. In the face of Waters, however, Munting is forced into an "excess of seeing," sympathizing with the dead Harrison. Confronted in the same evening with heteroglossic explanations for origins—as well as endings—of life, Munting contemplates an intelligent cause behind both: that perhaps God caused life and Latham caused death. *The Documents in the Case* thus displays an intelligent design about situated investigations of intelligent design. As a creation about the possibility of Creation, its writing performance anticipates *The Mind of the Maker*.

After *Documents in the Case*, Sayers published six more Peter Wimsey novels. However, it was not until *Gaudy Night*, "her most mature novel,"[25] that Sayers self-consciously comments upon the need for dialogized heteroglossia. She shows Harriet Vane coming to grips with the fact that Wilfred, the protagonist of her detective fiction, is monologic. Harriet thus spends four days "altering and humanizing Wilfred" by giving him plural sensibilities. She compares her new creation to the ivory chess pieces given her by Peter, made up of concentrically rotating globes within globes:

> Wilfred was becoming like one of those coveted ivory chessmen. You probed into his interior and discovered an intricate and delicate carved sphere of sensibilities, and, as you turned it in your fingers, you found another inside that, and within that, another again. (*Gaudy* 409)

The Hero, Wilfred, was becoming heteroglossic in the hands of his Author, Harriet, guaranteeing greater power to the novel that contained him.

There is no doubt that the novels which contained Harriet—*Strong Poison*, *Have His Carcase*, and *Gaudy Night*—became increasingly powerful as well.[26] However, once Sayers had performed in a new theater of representation—creating plays about the Creator of life—she did not return to detective fiction. Her dismissal of the genre in *The Mind of the Maker* reflects her awareness that theater can better dialogize heteroglossia. For, to make drama work, a playwright must create a polyphony of subject positions all of which, when incarnated on the stage, must seem truly human despite their differing perspectives. Furthermore, Sayers recognized that these varied perspectives reflect the heteroglossia within the successful playwright:

> [H]is center of consciousness shifts as he goes, so that in writing down John's lines he seems to view the stage from John's point of view, while in writing Mary's reply he views it from Mary's point of view. (*Mind* 55)

She gives the example of Shakespeare, who "is Iago as well as Othello;...because each is to some extent an expression of himself" (51). In *The Mind of the Maker*, then, Sayers explicitly eschews monologism: "if a character becomes *merely* a mouthpiece of the author, he ceases to be a character, and is no longer a living creation. Still more, if *all* the characters speak with their author's voice, the whole work loses its reality, and with it, its power" (52). Bakhtin calls this problem an "expressive" imbalance: "the author puts his own ideas directly into the mouth of the hero from the standpoint of their theoretical or ethical (political, social) validity, in order to convince us of their truth and in order to propagandize them" (*Author* 10). As Sayers notes, the problem with propaganda is its lack of diversity: "The Energy is active only in one part of the whole, and in consequence the wholeness is destroyed and the Power diminished" (*Mind* 53). Wholeness, for both Bakhtin and Sayers, is plural: a dialogism modeled on the separate but interdependent persons of the trinity.

The trinity was conceptualized differently within the Russian Orthodoxy of Bakhtin's homeland than it was in the Roman tradition with which Sayers was familiar. According to Catherine Mowry LaCugna, the trinitarian metaphysic of the West was situated upon the concept of one "substance" in three forms, thus presenting God as "something in and of itself." In contrast, the trinitarian theology of the East emphasizes that "communion underlies being." Hence "personhood," like that of the trinitarian God, implies "someone toward another."[27] As we have seen, "someone toward another" is fundamental to Bakhtin's moral philosophy as well as his literary theory; hence, a Russian Orthodox view of the Trinity may inform many of his writing performances. Significantly, Sayers got interested in Eastern Orthodoxy several years after *The Mind of the Maker* was published. In 1944, she wrote Father Kelly, the man who encouraged her to develop her trinitarian aesthetic, "I have just been reading Dr Jernov's *The Church of the Eastern Christians*, which was so attractive that I almost wanted to rush out and get converted to Orthodoxy immediately" (*Ltrs* 3. 473, nt. 1). Sayers's reading of Berdyaev, who sprouts up in various of her writings and is harvested in *The Mind of the Maker*, most likely planted the seeds of this attraction.

Influenced, it would seem, by Russian Orthodox views of creativity, both Sayers and Bakhtin were able to resist modernist perceptions of art, developing literary theories based on premodern doctrines of trinitarian creation. David Lyle Jeffrey notes that premodern hermeneutics did not locate "meaning" in the "plenitude" of language, but in a dynamic interrelatedness of author, text, and reader. He cites Wyclif in the fourteenth century, whose

> point is not only that for the text to be text-as-discourse it must be rooted in a dynamic relationship in which reader and text are brought together, but, more importantly, that an open disposition toward the author-in-the-text most readily permits the process of reading to create its fullness of possibility.[28]

Significantly, this premodern perception parallels postmodern reading strategies, as when Derek Attridge suggests that "it is not the text 'itself' but my singular and active relation" as reader that makes it significant: "a full response to the otherness of the text includes an awareness of, a respect for, and . . . a taking of responsibility for, the creativity of its author."[29] In his 1999 essay, Attridge often sounds like Sayers and Bakhtin, both of whom believed that a text was "actively produced, both by the author and by the contemplator" (*Author* 67).[30] Attridge discusses the engagement of the reader in ethical terms: "it will involve a suspension of my habits, a willingness to rethink old positions in order to apprehend the text's inaugural power." Writing performances thus elicit new performances, "as if the work being read demanded a new work in response" (Attridge 25). Sayers, who would have delighted in Attridge's employment of the word "work," makes a very similar point:

> When the writer's Idea is revealed or incarnate by his Energy, then, and only then, can his Power work on the world. . . . But once the Idea has entered into other minds, it will tend to reincarnate itself there with ever-increasing Energy and ever-increasing Power. It may for some time incarnate itself only in more words, more books, more speeches; but the day comes when it incarnates itself in actions, and this is its day of judgment. (*Mind* 111)

Not only do "actions" become incarnated through a type of rewriting—performance in new form[31]—but there can be multiple rewritings of any writing performance. Attridge suggests,

> However old the text, however familiar to me, it can always strike me with the force of novelty if, by means of a creative reading that strives to respond fully to the singularity of the work in a new time and place, I open myself to its potential challenge. (26)

Sayers advocates a similar openness, so that writing can perform its Power: "Sometimes we feel that a critic or student of a man's work has 'read into it' a good deal more than the first writer 'meant.' This is, perhaps, to have a rather confined apprehension of the unity and diversity of the Power" (*Mind* 121). Instead, asserts Sayers, new and varied readings reenergize a text: "your response to it will bring forth further power, according to your own capacity and energy" (122). She experienced this Power most exuberantly through her reading of Dante, telling Charles Williams that "one may properly 'read into' his writing anything that the words will carry of suggestion and human experience" (*Ltrs* 3. 108). Sayers, as reader, re-performed Dante's writing, just as she believed Dante re-performed the writings of

Thomas Aquinas,[32] for all readings performed with "power" become re-writing performances. "In the reading of literature," as Attridge puts it, "meaning is simultaneously formed and performed." Furthermore, this performance is relational: "the other" encountered in reading "is a relation or relating rather than an object" (Attridge 27, 25).

Like the otherness inherent in the Trinity itself, otherness became, for Sayers, necessary to any successful writing performance, reflecting what has been called an author's "realisation of the other in the self." Though the words sound straight out of postmodern theory, they are from Sayers, who believed that an author's creation parallels that of God, which is "given a genuine, though contingent, real existence of its own, so that it can stand over against Him and know Him as its real Other."[33] Because God, through Creation, performs the Other, creative writing, at its best, performs otherness.

Chapter 5

The Performing Word: Sayers's Unorthodox Orthodoxy

Now, in all myth-founded religions—and Christianity is no exception—the act of worship tends to issue in a ritual drama: a rite, that is, in which the events of the myth are dramatically re-enacted. The central drama of Christian worship is the rite of the Mass.

—Sayers, "Sacred Plays"

When *The Zeal of Thy House* went on tour in 1938, Sayers became the focus of increased media attention. In response to one interview, she composed "The Greatest Drama Ever Staged," which the *Sunday Times* published on April 3. As she told one correspondent, "That did it. Apparently the spectacle of a middle-aged female detective-novelist admitting publicly that the judicial murder of God might compete in interest with the Corpse in the Coal-Hole was the sensation for which the Christian world was waiting" (*Ltrs* 4. 139). The performance of *Zeal*, then, radically changed Sayers's life. Asked to write a second Canterbury work, *The Devil to Pay*, and a radio play about the birth of Jesus, *He That Should Come* (both published in 1939), she received an increasing number of invitations to speak and write on theological topics.

Almost against her will, Sayers started performing in a new theater of representation, becoming an apologist for Christianity at a moment in history when the intellectual respectability of Christianity was at its nadir.[1] High Modernism valorized "enlightened" individuals who had surmounted the anachronistic language of religion, asserting that truth was attained through the logic of science while meaning came through the explorations of art. In response, Fundamentalist Christians retreated into anti-intellectualism, saying that the Bible (the King James Version) was the sole scientific and artistic document they needed. Their escapism was exacerbated by the hegemony of logical positivism in the 1920s and 1930s, wherein only the language of empirically verifiable events was considered legitimate.[2] Sayers herself admits that, because she was "not religious by nature," she could have easily become a Logical Positivist (*Ltrs* 4. 139).

Fundamentalist Christians reacted to logical positivism using arguments that were more performative than they realized, for their vociferous writings unwittingly performed according to the scripts of their foes. Importing positivist assumptions into their reading of Scripture, they established the Bible as a collection of unassailable empirical statements, as if, in agreement with positivists, spiritual truths evaporate when scientific accuracy cannot be proven. In what Sayers describes as a "singular piece of idolatry,"[3] Fundamentalists were treating verses of the King James Bible like pieces of the cross, totemic signifiers that contained what Jacques Derrida has called the "Transcendental Signified": the full presence of truths to which they refer. Due to the stridency of this totemism, any Christian apologist, no matter how articulate, was denigrated by the secular intelligentsia as either "a mountebank, a reactionary, . . . an escapist, an obstructionist," or "an enemy of society," to quote Sayers (*Search* 69).

The modernist contempt for faith vocabularies perhaps reached its apogee in *The Emperor's Clothes* by Kathleen Nott (1953). Reprinted several times, Nott's book disparages "poets and critics who have attached themselves more or less firmly to the cause of dogmatic theology," asserting that they are "engaged in the amputation and perversion of knowledge."[4] She is especially contemptuous of C. S. Lewis and Dorothy L. Sayers, calling them "braver and stupider than many of their orthodox literary fellows," because of their "tub-thumping" popularizing of the faith (68, 43). With modernist condescension, Nott contrasts Christian dogmatism with the "open-minded inquiry" of science (298):

> If men learn to think consistently in the direction that leads to aeroplanes, they will also think in the direction away from authoritarianism and therefore theology. Moreover they will learn to think in a direction . . . toward psychology and anthropology and away from Sin and the Church. (16)

Nott seems to echo Rudolf Bultmann's famous "demythologizing" statement published the same year as *The Mind of the Maker*: "We cannot use electric lights and radios and, in the event of illness, avail ourselves of modern medical and clinical means and at the same time believe in the spirit and wonder world of the New Testament."[5]

Because Sayers can offer no "scientific examination" that establishes "proof" of the Trinity, Nott reviles *The Mind of the Maker*, extending her criticism to Sayers's view that language "always is metaphorical":

> [H]er attack on the scientists for trying to control the metaphorical twists of language . . . is purely negative. She does not seem to understand that a striving for the utmost precision may be a duty, and a positive one in scientific work. (Nott 286)

Ironically, in her attempt to condemn Sayers's Christian dogma, Nott dogmatically mystifies science, perpetuating what Richard Rorty calls "a picture of the scientist as a sort of priest."[6] As Nicholas Wolterstorff argues in his invaluable work *Reason Within the Bounds of Religion*,

> [T]he faith which the positivists displayed in natural science was not itself arrived at scientifically. On the contrary, it resembles in striking ways the confidence of the Congregation of the Inquisition in the veracity of Holy Scripture. (20)

Though Nott claimed that she did not deify science, she maintained a profound belief in the power of logic to innocently access truth. As late as 1982 she writes, "I claim only that the way we think logically, consistently and realistically is on a continuous line from commonsense immediacy to the scientific method of inquiry," and she asserts that the scientific response to "relevant evidence" is both "informed and unprejudiced."[7] By arguing for unprejudiced reason, Nott ironically reveals her own positivistic prejudices, appearing uninformed about the situatedness of all knowledge, including scientific logic. As Thomas S. Kuhn argued in *The Structure of Scientific Revolutions* twenty years earlier,

> No natural history can be interpreted in the absence of at least some implicit body of intertwined theoretical and methodological belief that permits selection, evaluation, and criticism. If that body of belief is not already implicit in the collection of facts—in which case more than "mere facts" are at hand—it must be externally supplied, perhaps by a current metaphysic, by another science, or by personal and historical accident. (16–17)

Similar to Kuhn, Sayers asserted that knowledge is often embedded in "preconceptions so ingrained in us that we do not realize our dependence upon them until we hear them challenged." She distinguishes such "assumptions" from "the doctrines we preach," implying that thinking Christians, unlike many positivists, are at least self-conscious about their commitment to belief.[8]

Sayers's faith-filled resistance to Nott-like modernism anticipates the postmodern recognition that the scientist has no better access to "the real" than does the Christian, for all truth is always already embedded in language. Indeed, Sayers sounds decades ahead of her time in her view that the sciences, in their attempt to "evolve precise languages and to avoid metaphor and analogy are in fact struggling to escape from our necessary conditions of truthful expression" (*Clothes* 294). With these words Nott summarizes Sayers's position in order to dismiss it: "[Sayers] tells us that we must think by analogy or refrain from thinking at all, and that thinking about God is merely no exception to this rule" (295). Nott's dismissal ironically foregrounds Sayers's prescience, seen when, nearly forty years later, Richard Rorty writes, "speaking a language . . . is not a trait a man can lose while retaining the power of thought. So there is no chance that someone can take up a vantage point for comparing conceptual schemes by temporarily shedding his own."[9] All thought arises from faith—faith in a conceptual scheme that is performed by language. Sayers says as much in a letter to a skeptic:

> One act of faith must, indeed, be made before one can accept Christianity: one must be prepared to believe that the universe is rational, and that (consequently) human reason is valid so far as it goes. But that is an act of faith which we have to make in order to think about anything at all . . . . Admittedly, we cannot prove that the universe is rational; for the only instrument by which we can prove anything is reason, and we have to assume the rationality of things before we can trust or use our reason. (*Ltrs* 2. 401)

Sayers has described a paradox often noted by poststructuralists, wherein the "truth" of a thing is dependent upon a vocabulary that must presuppose the "truth" that it is trying to prove. She implies that if "rationality" itself must be taken on faith, one may as well believe in a God that gives rationality meaning, a God who created a world in which justice and goodness are reasonable. In her broadcast talk "The Religions Behind the Nation," she asserts,

> [A]t the very basis of our thought and behaviour there lies a pair of assumptions which are wholly religious—which reason cannot prove

> and for which science can offer no evidence. We assume that *both* our conception of the good *and* our human reason are really valid. (70)

Christian dogma, then, grounds these assumptions in the nature of God, claiming that humanity's "persistent belief in goodness and reason were [*sic*] justified; that such was the nature of God and the true nature of man" (73).

Some readers might assume that Sayers simply derogates modernist scientism in order to position her own belief system as an unassailable "myth" that subsumes all the rest—what postmodern theorists call a "metanarrative." When Sayers employs the word "myth" to describe the Christian story, as in the epigraph that begins this chapter, she does not imply that Christianity is the performance of fanciful falsehood. She uses "myth" in its technical sense: a metanarrative that makes meaning of existence by explaining humanity's relationship to transcendent and metaphysical realities. Sayers, however, does not treat the Christian "myth" as an unquestionable metanarrative. She questions it repeatedly by assailing the language through which it is mediated, thus garnering accusations of blasphemy for several of her unconventional writing performances. Her second book of poetry, *Catholic Tales and Christian Songs* (1918), created controversy when a prominent Catholic poet, Theodore Maynard, published a negative review in *The New Witness.* A Sayers friend responded by writing two disingenuous letters to the editor under different pseudonyms, a move orchestrated by Sayers in order to generate curiosity and thus induce sales of her work: a writing performance in several senses of the term!

Maynard's accusation, that many of Sayers's poems align Christianity with pagan myths, is justifiable. While "Christus Dionysus" identifies Jesus with the Greek God of wine and fertility, "Dead Pan" parallels him with the goat-footed Roman god as well as with the Egyptian god of the dead, Osiris.[10] "Rex Doloris" employs sexual innuendo as the Greek goddess Persephone longs to be reunited with Christ, who is imaged as the "great King of Grief" who "hath reft me, ravished me" and who "shall have me at His will" (30, 31). Erotic overtones also can be heard in "Desdichado," which paints Christ as a "rascal fiddler" who sings "Lady, lady, will you come away with Me, / Lie among the bracken and break the barley bread?" (7). This poem, like several others, employs imagery reminiscent of Medieval Romance; "White Magic" calls Jesus the "Wizard-Man from Nazareth," and "The Wizard's Pupil" turns the Bible into a Wizard's book of spells that "wakes the dead" (25, 14). Fundamentalist Christians in the twenty-first century who protest the sorcery in Tolkien's *The Lord of the Rings* trilogy would be appalled.

Ironically, even though Sayers's metaphors and analogies in *Catholic Tales and Christian Songs* were unusual enough to generate controversy in 1918 (with a little help from her friends), they were not unprecedented. The parallel between Pan and Jesus goes back to Plutarch and is employed by Milton in his poem "On the Morning of Christ's Nativity." Medieval quest narratives were imbued with Christian symbolism, as was the mysticism of alchemical wizardry. And one need only look at Donne's "Holy Sonnets" to see sexual conceits transmitting Christian doctrine.

Nevertheless, Sayers knew she might shock her readers, as is implied by the poem that begins *Catholic Tales and Christian Songs*, in which she asks "pardon" of Jesus (3). Aware that some might see her poetry as a slap in Jesus's face, Sayers signals that her ultimate goal in the book is to serve his cause. She does so by undermining culturally constructed religious signifiers that have become as holy as the signified reality toward which they gesture. This is especially apparent in the longest piece in *Catholic Tales and Christian Songs*, a closet drama called "The Mocking of Christ," which illustrates various ways that Christians have codified Christ, appropriating him as an emblem to support their own interests. True to many of Sayers's writing performances, the play uses convention to unconventional ends; the convention here comes from Medieval Corpus Christi plays, which contain extensive scenes of Christ's torture: the "buffeting," the "scourging" and, of course, the crucifixion itself. Sayers applies these conventions to the history of Christianity, suggesting that the church repeatedly tortures Jesus through its self-serving agendas.

"The Mocking of Christ" begins with a statement of the "Persona Dei," who is then declared "It" for a game of tag by Roman Soldiers (43, 44). Next enters the Pope who hands Jesus a robe, crown, and keys, thus treating him not much differently than do the soldiers before his crucifixion. An Emperor joins the Pope, giving Jesus a "temporal crown, sword and sceptre" in illustration of the political power that the Roman Church wielded after the conversion of Constantine: a mockery of all that Jesus stood for while on earth. A King then enters who begins to fight with the Pope, presumably to wrestle power away from him; however, their "blows miss each other and fall upon Christ," as though to say that the church's wrangling for political hegemony scourges Jesus all over again. The King wins out, saying "Who keeps quiet and serves the King / Can't go wrong in anything"; he "binds a charter upon Christ's mouth," alluding to the way in which the "Divine Right" of kings silenced the revolutionary teachings of Christ. Next Sayers has a Protestant preacher place "a black gown and a shovel hat" upon Jesus and hand him a Bible while saying "You still speak by the Book / Never say a word / But what we've always heard." This reference to the Protestant idolizing of traditional interpretations of the Bible is then displaced by the entrance of a character who gives Christ a bishop's apron and gaiters, in deference to Victorian "good taste" and respectability. He is followed by one curate who sets a cup of tea in Christ's hand, then by another who gives him a cricket-bat and pads, slapping him on the back like a sports buddy (45–46).

These nonviolent treatments of Jesus are perhaps most chilling of all, for they imply a cultural appropriation of the Word that elevates decorous graciousness over salvific grace, propriety over prayer, as though to say Christ sanctions the status quo. Indeed, societal interests once again reign as a "Respectable Gentleman" winds Jesus in red tape and calls him "Chairman of Committee," thus turning him into a mere figurehead for bureaucratic functions. After two patriots from different nations both claim that Jesus endorses their opposing national causes, a sentimentalist "parts Christ's hair in the middle," grooming a "Jesus mild and meek" who will threaten nothing. Finally, a "Higher Critic" coordinates a parade of numerous figures from history and mythology, all of whom claim Jesus to be either a manifestation of themselves or a fulfillment of their philosophies (47, 48). Jesus thus becomes a construction of various cultures: the sum of multiple writing performances.

Sayers's own writing performance ends as it begins, with a statement from the "Persona Dei," who is now ready to be hung on the cross:

> Now that I must come to die
> Nought is left of Me, save I
> Discrowned, stript, alone. (52)

Sayers quite clearly indicts multiple manifestations of Christianity for stripping Jesus of his redemptive word and placing upon him culturally validated emperor's new clothes. Hence Sayers attempts to redress Jesus in the other poems of *Catholic Tales and Christian Songs*, putting him on a stage containing new props of language, props so abnormal that they sound blasphemous.

Richard Rorty distinguishes between "abnormal" and "normal" language in *Philosophy and the Mirror of Nature*, explaining that the abnormal occurs when someone sets aside "an agreed-upon set of conventions" (320). When Sayers sets aside the normal "stained-glass-window decorum with which the tale [of Jesus] is usually presented to us,"[11] she incarnates what Rorty, borrowing from Harold Bloom, calls "the vanguard of the species": the "strong poet" (*CIS* 20). For Rorty,

> The line between weakness and strength is . . . the line between using language which is familiar and universal and producing language which, though initially unfamiliar and idiosyncratic, somehow makes tangible the blind impress all one's behavings bear. (*CIS* 28–29)

Sayers becomes a "strong poet" when she makes tangible the stained-glass blindness of Christians to their conventional behavings—behavings that are the unthinking performance of culturally constructed scripts. Like Rorty, Sayers believes that the "function of imaginative speech is not to prove, but to create—to discover new similarities and to arrange them to form new unities."[12] The "new" arises from "unfamiliar and idiosyncratic" language that forces people to consider the contingency of their vocabularies, vocabularies which, for Sayers, obscure the "abnormal" implications of the Incarnation.

A tension between normal and abnormal language informs Sayers's first dramatic writing performance to follow *The Zeal of Thy House*: a radio play about the birth of Jesus. In *He That Should Come*, Sayers aligns protection of "the normal" with a Pharisee lodging in an overcrowded inn on the night of Jesus's birth. He says to a Jewish Gentleman, dressed in Roman fashion, "God deliver us from the fashions of Rome—where they teach our Hebrew youth to sneer at God's word and bind a foreign yoke and a pagan custom upon our necks in flat defiance of the Law of Moses" (154). Though Sayers was surely as disturbed by the Roman oppression of the Jews in the first century as she was about the Nazi oppression in 1939,[13] she quite clearly shows that this Pharisee is not dismayed by the loss of freedom so much as he is by the loss of traditional performances. Indeed, in the same scene, she has the Pharisee compliment a Merchant for using the "old-fashioned" town name "Joppa"—language that is familiar—rather than the Romanized "Caesarea." The Merchant responds with, "A heathen name, of course, but what's in a name?" (156). Sayers, of course, realized that a great deal was in a name; normal language has the power to stigmatize and stultify as it perpetuates customary practices, but an abnormal name has the power to stretch perceptions.

As we have seen in *Catholic Tales and Christian Songs*, Sayers believed that the employment of "heathen" or "pagan" language can infuse flaccid religious verities with new life. For her, the intellectual energy arising from shock was far preferable to the stagnate piety that comes from unreflectively clinging to the writing performances of religious convention. As she implies in her essay "The Greatest Drama Ever Staged," Christians who normalize Christianity fail to acknowledge the exceedingly shocking performance of the one after whom they are named:

> The people who hanged Christ never, to do them justice, accused Him of being a bore; on the contrary, they thought Him too dynamic to be safe. It has been left for later generations to muffle up that shattering personality and surround Him with an atmosphere of tedium. We have very efficiently pared the claws of the Lion of Judah, certified Him "meek and mild," and recommended Him as a fitting household pet for pale curates and pious old ladies. (*Creed* 6)

In respect for that shattering personality, Sayers chose to supplant normal with abnormal writing performances. She thus became like Rorty's "strong poet," writing performances "dominated by metaphors of making rather than finding, of diversification and novelty rather than convergence to the antecedently present" (*CIS* 77). The "strong poet," in other words, has "the mind of the maker." As Sayers puts it in a talk she delivered in 1942, "Poets create, we may say, by building up new images, new intellectual concepts, new worlds, if you like, to form new consistent wholes, new unities out of diversity."[14]

It is not surprising, then, that Sayers makes some "heathen" characters in *He That Should Come* more admirable than the pious Pharisee, for they are open to new words, as literalized in the birth of the new Word among them. A Roman Centurion, who praises the tolerance of Rome for "all sorts of odd foreign deities" (161), is the one who compassionately persuades the innkeepers to give shelter to Mary and Joseph, asking the landlord, "How about the stables? Is there any *room there*?" In contrast, the Pharisee makes no room for those outside his worldview: "*What room* will there be for such as you in the great day of redemption when the Lord's Messiah comes?" (150, 158; emphasis mine). Of course, Sayers implies that the Pharisee is so obsessed with a traditional interpretation of the Messiah that he fails to see the new Word when it appears before his very eyes. In contrast, the "Jewish Gentleman" who dresses in the new Roman fashion turns out to be Joseph of Arimathea, who will, in an act of love and respect, provide room in his tomb for the crucified Jesus. Sayers implies that the Kingdom of God is established not through the idolatrous maintenance of tradition, not through preserving conventional writing performances, but through love—something she applies even to her relationship with the author of *The Emperor's Clothes*:

> If Miss Nott were here now, she and I could establish the Kingdom of Heaven between ourselves immediately—that is, we could *if* we could. It is quite simple: she has only to love me as well as she loves herself, and I have only to love her as well as I love myself, and there is the Kingdom.

Sayers, however, goes on to acknowledge that without the power of the new Word, the dogma that appeared in a manger, such love is exceedingly difficult: "That is the trouble with the Christian graces—that without Grace they are impossible."[15]

While some may choose to interpret the tension between love and law within *He That Should Come* as an ironic example of Sayers's own blindness to the performativity of anti-Semitism in her day, Sayers clearly shows that the Pharisee's problem is not his faith position so much as his inflexible arrogance. He responds to the Jewish Gentleman's praise of Roman theaters with, "Nothing would induce me to look at them. Play-acting and wild-beast shows are an abomination in the sight of the Lord. Immoral, irreligious, and thoroughly un-Jewish" (157). Significantly, Sayers gives the same opinion to one of the play's shepherds; however, this Jew who humbly presents himself to Jesus also speaks humbly:

> And there's a young chap I know, that's employed in the theatre, as they call it, at Jerusalem, says the goings-on there is something shocking—men dressed up like women with masks on, acting heathen pieces full of smut and nastiness.... It ain't right, to our way of thinking. (167)

While the Jewish Shepherd acknowledges the situatedness of his ethic—"our way of thinking"—the Pharisee claims to know the mind of God.

The Pharisee's problem is not his Jewishness but his lack of epistemological humility. Indeed, a dozen years later Sayers was to give a similar personality to a Christian character, Arius, in *The Emperor Constantine* (1951), her turgid play about the Nicene Creed. Arius is shown to have Biblical justification for his view that Jesus, though one with God, was God's creation. However, as Janice Brown points out, Sayers has made him "arrogant to the point of insolence,"[16] so that it is his supercilious attitude, not his doctrine, that makes him distasteful.

When he is asked to present his view at the Council of Nicaea, he states,

> I shall be only too delighted—both to defend my name from my traducers and to defend my doctrine against those who are too illiterate to understand Greek, and too indolent to study the Scriptures. . . . Our Latin friends who have no definite article in their woolly language may be excused for woolly thinking; but for those who speak Greek there is no excuse. (143)

Like her Pharisee in *He That Should Come*, Sayers's Arius assumes his language is superior to the fashion that comes out of Rome, brought by "our Latin friends." This is significant in light of the fact that Sayers herself valued the sacramental Christianity of the Roman church, identifying herself as an "Anglo-Catholic," a position considered "daring and controversial" in her day.[17] Even in her orthodoxy, then, Sayers was abnormal.

Well educated in Catholic doctrine, Sayers cites famous church heresies throughout various letters and essays.[18] One might wonder, then, how her advocacy of orthodox Christianity and its creeds harmonizes with her challenges to "the old familiar obsolete phrases [that] have been sterilized by pious associations" (*Ltrs* 2. 357). Sayers, I would like to suggest, exemplifies what Dale Irvin calls "Christian Traditioning," a practice of making "present the historical past as memory and identity." As a "constructive activity" that establishes the relevance of ancient faith to contemporary existence, this practice keeps a tradition alive: "Where there is a living sense of the historical past or a historical consciousness, there is an effective tradition and a moment of traditioning" as the convention is expressed in new ways.[19] Sayers endorses "traditioning" in her 1941 address to the Archbishop of York's Conference at Malvern, defining "a Church" as a body of humans "organized within a living tradition whose essence persists unchanged while its expressions continually develop."[20] And five years later she was to illustrate Irvin's sense of "traditioning" with Thomas Aquinas, "a rather dangerous 'modern' who had incorporated a pagan scheme of philosophy into Christian dogmatic teaching, somewhat to the scandal and alarm of the old-fashioned orthodox" (*Further* 38).

Sayers anticipates elements within the postmodern theology of "Radical Orthodoxy," which seeks to recuperate Aquinas as presciently recognizing "the situatedness of our manner of knowing," as Catherine Pickstock puts it: "for Aquinas truth is not at all a matter of detached abstraction, but rather of the specific entry into our minds of certain contingent features and events. It is a matter of the conjoining of ourselves with others in time and space."[21] Aquinas therefore exemplifies the traditioning that "radically orthodox" theologians want to offer as the vocation for all Christians:

> Tradition, the *traditio* or "handing over," is not simply something which is handed over to me, but rather something over to which I am first delivered, am "proper to." In this sense "I" am constitutive for the tradition as being in an intimate dialogue with it: I am the potential horizon of its being made actual, its realization.[22]

Christian tradition is "made actual" through the performance of Christian believers who act upon traditional writings; the "handing over" of these writings is therefore a writing performance.

According to Dale Irvin, the Greek word for "handing over"—*paradidomi*—is used in the New Testament to describe not only the "handing over" of Christ's words to his followers, but also the "handing over" of Christ to his enemies (40). These two senses come together in Paul's first letter to the Corinthians: "For I received from the Lord what I also delivered (*paredoka*) to you, that the Lord Jesus on the night when he was betrayed (*paredideto*) took bread..." (11:23). The etymological connection between delivery and betrayal leads Irvin to an interesting conclusion:

> In every act of authentic traditioning there remains something of an act of treason, otherwise it would not be an authentic act of handing over, of change. Without a bit of treason performed in the act of handing over, the tradition remains inseparably bound to the world in which it was formed, hence not only irrelevant but incomprehensible. (41)

Significantly, Dorothy L. Sayers, who knew that "tradition . . . means the 'handing over'" (*Search* 79), was accused of treason for writing performances about the life of Jesus. Her wartime radio play cycle, *The Man Born to Be King*, elicited accusations that "Singapore fell because these plays were broadcast." Protesters therefore "appealed for them to be taken off [the air] before a like fate came to Australia."[23] Sayers's abnormal writing performance was interpreted as an act of "handing over" countries to the enemy.

Sayers wrote *The Man Born to Be King* at the request of Dr. James Welch, director of Religious Broadcasting for B.B.C. radio. By December of 1941 the series was ready to be announced, and Welch set up a press conference to advertise *The Man Born to Be King*, at which Sayers read some dialogue from plays she had already finished. When the media sensationalized her abnormal language, a censorship campaign was mounted by the Protestant Truth Society and the Lord's Day Observance Society (LDOS), with petitions seeking to ban *Man Born to Be King* going to both the prime minister and the Archbishop of Canterbury. In the protesters' minds, Sayers's traditioning of Christian dogma was a betrayal—a "handing over"—of the "Authorized Version" of truth.

What especially upset the LDOS was Sayers's "use of many modern slang terms," for it meant "a spoliation of the beautiful language of the Holy Scriptures which have been given by inspiration of the Holy Spirit."[24] People were horrified that some of the slang was *American*—a far cry from the King James English that the LDOS and Protestant Truth Society were seeking to protect—as though to say Jesus actually spoke in Jacobean English. Ironically, of course, a good portion of the King James Version of the Bible is based on the "traditioning" of William Tyndale, who was interdicted a century earlier for seeking to render New Testament Latin into vernacular English and whose translations were suppressed by the English church. Language reverenced by Christians in Sayers's day had once been declared heretical by equally earnest Christians four hundred years earlier. The "handing over" of the Bible by Sayers, like that by Tyndale, was considered a betrayal of sacred writing performances.

Sayers believed, however, that when Christians mystify the language of the Bible, they lose the ability to identify with the performances it describes. She parodies the resulting perception—that "the characters in the Bible are all sacred"—with satirical words:

> Christ wasn't born into history—He was born into the Bible (Authorized Version)—a place where nobody makes love, or gets drunk, or cracks vulgar jokes, or talks slang, or cheats, or despises his neighbours, but only a few selected puppets make ritual gestures symbolical of the sins of humanity. No wonder the story makes so little impression on the common man. It seems to have taken place in a world quite different from our own—a world full of reverent people waiting about in polite attitudes for the fulfilment of prophesies. (Ltrs 2. 354)

Sayers decried this "prejudice which afflicts most Protestant communities," explaining that the protest to her plays "really represents a general fear and hatred of the human body."[25] For Sayers, the Incarnation—God taking on flesh—is more than a creed of the church; for her it makes the Christian story relevant to lived experience. She thus committed herself to making the characters in *Man Born to Be King* sound like humans with which her audience could identify: individuals—including Jesus—with physical needs and desires. As she tells one correspondent during the height of the scandal, her desire was to

> avoid any suggestion of easy slickness in dealing with the impact of the Divine upon the Human, however perfect the Human may be. Otherwise, the ordinary [audience member] feels merely: "What has all this to do with me? These people are not real men and women. Christ is not a Man—His temptations are a pretence, His works cost Him nothing, His sufferings are not the sufferings of real flesh."[26]

To make Christ's suffering seem real, Sayers repeatedly asserts in *Man Born to Be King* that Jesus, though one with God, did not always know the mind of God, thus explaining his prayer in the Garden of Gethsemane: "If it be possible, let this cup pass from me" (Matt. 26:39). In her notes to play nine, *The King's Supper*, she says of Jesus,

> In the Upper Room, the God-consciousness and the Man-consciousness are very closely fused, with the God-consciousness often coming to the top. In the Garden, it seems as though the Man-consciousness became permanently uppermost, and I think it must be played with a very strong and real sense of the human horror at physical death, of the failure of human companionship, and of the apparent vainness of prayer. (*King* 230)

Her phrase "permanently uppermost" implies that the Man-consciousness remained with Jesus through his death, such that his agonized prayer on the cross—"My God, my God, why hast Thou forsaken me?"—might reflect his human horror at physical death.

Sayers made the humanity of Jesus so vivid that it shocked her audiences. To do otherwise, implies Sayers in her introduction to *The Man Born to Be King*, is to sacrifice the message of the Gospel: "To make of His story something that could neither startle, nor shock, nor terrify, nor excite, nor inspire a living soul is to crucify the Son of God afresh and put Him to an open shame" (21). With these words we are also given an apologetic for the performance of pagan images and sexualized metaphors staged in *Catholic Tales and Christian Songs* twenty years earlier.

Sayers's commitment to breaking tradition out of conventional writing performances was inspired by the Roman Catholic detective novelist

G. K. Chesterton. In her preface to his play *The Surprise*, she employs an architectural metaphor to praise Chesterton's "traditioning":

> Like a beneficent bomb, he blew out of the Church a quantity of stained glass of a very poor period, and let in gusts of fresh air, in which the dead leaves of doctrine danced with all the energy and indecorum of Our Lady's Tumbler. No doubt, in her substance, the Church was, then as ever, glorious and immutable, but she was, at the beginning of the century, somewhat unfortunate in her accidental presentation, seeming to the man in the street, to offer little choice except between a somewhat complacent Protestant morality and a rather simpering Catholic piety. (5)

Sayers regards Christian tradition as glorious and immutable, and hence worth "handing over." However, "accidental"—meaning socially constructed—interpretations must be "handed over" to the iconoclast, who can deconstruct sanctified vocabularies about "morality" and "piety."

Sayers foregrounds the problematics of language in her second Canterbury play, *The Devil to Pay*, which takes an "abnormal" approach to the Faustus myth. Sayers has Mephistopheles mock Faustus's incantations with the statement "And now, sir, what can I do for you, to justify the expenditure of so many big words and this great exhibition of fi-fo-fum?" (33), as though to say it is not the power of language that conjured his presence. Indeed, at the end of the play Faustus acknowledges his self-serving pride by saying,

> Not in the words
> Is power, but in the faith of him that speaks,
> And in the person of the very Christ
> In Whom stands all the meaning of creation.
> Words? They are rags, tags, fluttering remnants blown
> Along the winds of fancy. (116)

For Sayers, even the words of the Bible, when decontextualized, become rags, tags, fluttering remnants too easily blown along the winds of fancy. As Faustus discovered, language cannot make present supernatural realities.

Eschewing what David Lyle Jeffrey describes as "a tendency among post-Reformation and evangelical Protestants to elevate the Bible to the status of a shibboleth, seeing it in effect as *autopistos*—drawing its authority from itself," Sayers recognized that Scripture is not self-interpreting.[27] When an American correspondent asked her to "write a book about the Scriptural sanction for the doctrine of the Trinity," she responded with an articulate problematizing of "Sola Scriptura" sensibilities:

> [W]here is your Scriptural authority for the Scriptures themselves? On what texts do you rely for the make-up of the Canon as we have it? Where, for example, does the Lord say that there are to be those four Gospels and no more? or that the *Revelation of Peter* and *The Shepherd of Hermas* are not authoritative—though the first was read in churches as early as the second century, and the second was included in the Codex Sinaiticus as late as the fourth century? The doctrine of the Trinity was worked out and formulated in the Church—the same Church that is the authority for the Canon itself. (*Ltrs* 2. 367)

In Sayers's mind, if the Biblical canon is contingent upon Church history, Christians should study the traditions and evaluate the proclamations of those who formulated the canon. As she tells another correspondent, "Many, indeed, of the most crude and erroneous ideas about doctrine (especially as regards redemption) are directly derived from the reading of the Bible without sufficient knowledge of its theological and historical background" (*Ltrs* 2. 288). "Traditioning," not proof-texting, is necessary to keep alive the truths of the Church.

Sayers demonstrated her own knowledge of Christianity's theological and historical background in *The Emperor Constantine*, a play commissioned by the Bishop of Colchester for a 1951 festival. The work dramatized not only Constantine's coming to faith, but also his involvement with the Council of Nicaea (325 A.D.), which established the doctrine that Jesus was "begotten, not made" and thus "of one being" with God. The play shows council participants laboriously considering every word of their creed, making sure it reflects their collaborative understanding of Christ's nature and mission. Sayers, however, does not make them protective of language for its own sake. As the Council considers the Greek word "homoöusios" ("consubstantial") as a way to establish, against Arius, Jesus's oneness with Creator God, it acknowledges that the word had once been "condemned" by the Eastern Church. Lactantius states, "A word that was open to misconstruction in the last century may be harmless and even salutary now" (117). Sayers once again implies that language is not an innocent container of meaning; it reflects cultural assumptions, performing differently as social values and practices change. In an article on Constantine's Nicaea written nearly fifty years later, Daniel Williams states, "Just as the Church is a historically developing organism whose life is the incarnated extension of Christ's incarnate life, so the orthodox or 'true' faith of any age is itself a construction with a particular context."[28]

Because makers of creeds construct meaning in the process of representing it, they must approach their task as a communal enterprise wherein the integrity of the work relies on majority assent. Sayers can therefore counter Kathleen Nott's attacks on Christian doctrine by drawing attention to the etymology of the word *dogma*: "an *opinion* endorsed by the whole body" of believers.[29] Significantly, whenever Sayers was asked her own opinion about theological issues, she demurred, repeatedly repudiating the "personal" in discussions of theology. As she tells one correspondent, "Nothing would induce me to 'set down my religious beliefs and convictions'. Setting down what I understand to be the Church's beliefs and convictions is a different matter" (*Ltrs* 2. 114). Rather than appeal to her own integrity, she appeals to the integrity of the work: "It isn't a question of agreeing with me—I have expressed no opinion. That is the opinion of the official Church, which you will find plainly stated in the Nicene Creed, whether or not you and I agree with it" (*Ltrs* 2. 180). Rejecting a modernist emphasis on the reasoning abilities of autonomous individuals who can access logically consistent texts, Sayers reflects a premodern epistemology. As Medieval scholar Joseph Huffman notes, for premodern Christians, "the ultimate, over-riding truth resided in God's righteous order in the community rather than in human forms such as written texts." Significantly, Huffman proceeds to demonstrate how premodern Christianity anticipates many postmodern paradigms.[30]

Because of her belief in the traditional "opinion" of the community—or "dogma"—Sayers strongly disliked the Evangelical emphasis on individualist piety and celebrity evangelism: a distaste manifest throughout her life. In 1927, she wrote her cousin Ivy about the three-year-old John Anthony: "if he shows any signs of being religious, I wash my hands of him" (*Ltrs* 1. 257). At age 45, having completed two and in the midst of writing a third play on Christian subjects, she told the Bishop of Nottingham, "I refuse to be a Christian evangelist" (*Ltrs* 2. 118). Several years later she was to tell a reader of *The Mind of the Maker*,

> I am weary of this evil and adulterous generation, with its monstrous deification of insignificant personalities. If a thing is not true in itself, the fact that I say it will not make it any truer; nor is it any addition to God that a popular novelist should be so obliging as to approve of Him. (*Ltrs* 2. 316)

But how could she know Christian dogma was "true in itself"? Sayers's answer is suggested in *The Mind of the Maker*, where she employs an architectural metaphor in order to parallel an artist's "truth" with that of God:

> [N]o considerations of false reverence should prevent us from subjecting the incarnations of creators to the severest tests of examination. It is right that they should be pulled about and subjected to the most searching kind of inquiry. If the structure is truly knit, it will stand any strain, and prove its truth by its toughness. Pious worshipers, whether of mortal or immortal artists, do their deities little honor by treating their incarnations as something too sacred for rough handling; they only succeed in betraying a fear lest the structure should prove flimsy or false. (92)

In other words, if the structure or "intellectual pattern" of Christianity coheres through the ages, despite self-serving appropriations made by its followers (as illustrated in "The Mocking of Christ"), it must be "true."[31] And of this Sayers was so adamantly convinced that she felt free to challenge the language by which the coherent structure is mediated. As she tells one correspondent, "What I say is what the Church says—only the language is different" (*Ltrs* 4. 140). She changed the language in order to portray Church dogma as spectacularly exciting:

> It is the dogma that is the drama—not beautiful phrases, nor comforting sentiments, nor vague aspirations to loving-kindness and uplift, nor the promise of something nice after death—but the terrifying assertion that the same God who made the world lived in the world and passed through the grave and gate of death.[32]

Sayers earnestly believed, however, that this terrifying assertion was more than a communally endorsed writing performance. What the church said—its opinion about salvation through the resurrected Jesus—was, to her, a universal truth. Though it may seem inconsistent for Sayers to emphasize both the particularity of a community's opinion and the universality of that opinion, she would see it as no less inconsistent than those who make an absolutist statement by denying the existence of absolutes.

In *Contingency, Irony, and Solidarity* Rorty asserts that, because knowledge is contingent upon language, "truth" can only be assessed through the pragmatic outcomes of a community's structure of belief rather than by any correspondence to a reality that stands outside of culture. Rorty, in fact, denies that "Truth" is "out there" to be discovered and hence repeatedly puts "out there" in quotation marks throughout his book. In contrast, and in consonance with most people of her generation, Sayers believed that "Truth" is, indeed, "out there." However, unlike many intellectuals of her day, she would agree with Rorty that the understanding of "truth" comes by identification with a community rather than through insightful individuals who autonomously attain enlightenment through rational exploration. Rorty explains that "if languages are historical contingencies, rather than attempts to capture the true shape of the world or of the self, then to 'stand unflinchingly for one's moral convictions' is a matter of identifying oneself with such a contingency" (*CIS* 60). Sayers chose to stand unflinchingly for the contingent vocabulary of Christian dogma, believing that it provided the best explanation for both the world and the self:

> Christians have too weakly acquiesced in a vague religiosity and the worship of nothing-in-particular. We have been so anxious to avoid the charge of dogmatism and heresy-hunting that we have rather lost sight of the idea that Christianity is supposed to be an interpretation of the universe. (*Ltrs* 2. 158)

Sayers's reference to "interpretation" signals how she keeps her commitment to Christian dogma from turning into rabid dogmatism. Unflinchingly certain of God's work in and through the Christ-event on earth, she nevertheless knew that her performance as a Christian was ground in the contingency of language.

Sayers surely sympathized with the Greek Gentleman in *He That Should Come* when she has him say of the Pharisee, "What a very dogmatic person! It must be marvellous to feel so positive about everything. I never feel certain of anything" (173). Indeed, she turned down an honorary doctorate in divinity with the words, "I am never quite sure whether I really am [a Christian], or whether I have only fallen in love with an intellectual pattern" (*Ltrs* 2. 429). Significantly, eight years later, she gave Arius a similar pleasure in the intellectual pattern of Christianity. In *The Emperor Constantine*, she has a bishop say of Arius, "One does not feel in him any warm devotion to Our Lord—only an arid pleasure in a philosophical pattern" (114). Sayers, as well, did not exude warm devotion, often denouncing the "woolly emotionalism" of "religious sentiment" (*Ltrs* 2. 51, 262). Nevertheless, the epistemological humility expressed in her "I am never quite sure" contrasts with the presumptuous religious certitude born of snobbish intellectual autonomy she gives to Arius at the Council of Nicaea.

Sayers also strikes a blow at intellectual autonomy in *The Just Vengeance*, where she shows the performance of faith operating through the sacramental performativity of communitarian dogma. In the play, a recently deceased airman interacts with other dead people from his hometown of Lichfield (the location of the Cathedral that commissioned the 1946 production). As he struggles with the meaning of "justice," a "Recorder" prompts him to recite his Creed. He starts with "I believe in God," only to interrupt himself with

> No! no! no! What made me start off like that?
> I reacted automatically to the word "creed"—
> My personal creed is something totally different. (295)

By appealing to his "personal creed," the airman implies that he can autonomously attain truth. The Recorder, however, affirms how the community's discourse has molded his sensibilities, thus valorizing the performativity of creedal language within him:

> What is speaking in you is the voice of the city,
> The Church and household of Christ, your people and country
> From which you derive. Did you think you were unbegotten?
> Unfranchised? With no community and no past?
> Out of the darkness of your unconscious memory
> The stones of the city are crying out. (295)

Sayers uses the same metaphor she had employed a decade earlier in *The Zeal of Thy House* (59), alluding to Jesus's comment to the Pharisees during his Triumphal Entry into Jerusalem: if those who praised him were silenced, "the stones would immediately cry out" (Luke 19:40). In *The Just Vengeance*, she implies that, even though the airman seeks to silence a creed that he no longer personally believes, its truth-effects are nevertheless a part of the community by which his subjectivity is shaped and hence will inevitably cry out through him. Sayers even makes "sin" a part of the communal phenomenon that shapes the subjectivities of individuals. She has the airman make his confession by saying,

> Forgive me, I was wrong: we are victims together
> Or guilty together; if we have betrayed the future
> We will share the blame—(299)

Sayers made the same point, once again, in *The Zeal of Thy House.* The play's protagonist, William of Sens, like the airman in *Just Vengeance*, takes pride in his autonomy until confronted by the communal phenomenon that shapes him—literally, in the case of William, whose body becomes misshapen by an accident. When his lover, a rich widow, enters the cathedral, William's coworkers become distracted from their communal task: to inspect a rope that will suspend William from the top of an arch. Two men fail to see a flaw in the rope: the self-righteous monk Theodatus because he will not look at the lusty William and Ursula, closing his eyes as he says his prayers, and the vulgar workman Simon because he cannot take his eyes off the lovers as he sings a bawdy song. The sin is communal, a collaborative work of betrayal, as the Prior makes clear to Theodatus:

> This is thy sin: thou hast betrayed the work;
> Thou hast betrayed the Church; thou hast betrayed
> Christ, in the person of His fellow-man. (*Zeal* 75)

The betrayal, a "handing over" through the handling of rope, affects not just William, but also Christ and his church. Ironically, the unreligious Simon confesses his betrayal immediately, whereas the sanctimonious Theodatus at first denies complicity: "I had nothing to do with it" (74). Theodatus places personal piety above the integrity of the church's work—in several senses of the phrase. For the church is not only a work of architecture—Canterbury Cathedral—it is also the interdependent working of a community, all of whom must confess sin, because all are complicit with "The Fall"—literalized in *Zeal* through William's fall from the arch.

Just as sin is communal, so is its redemption, symbolized, for Sayers and her High Church peers, through the Sacrament of Communion. In a 1942 talk delivered to the North London Presbyterian Fellowship of Youth, Sayers explained that, for the Anglo-Catholic, "the act is never *wholly* individual. Each communicant makes and partakes of the sacrifice in the name of the *whole* Church."[33] The "sacrifice" is that of Christ, "*acted* and *performed*" by each communicant during Mass: "It is the reenacting upon the stage of the world of the great drama of the Passion—a drama acted in His name by priest and people" (Worship 43). Sayers's emphasis on performativity, wherein the "acting of the thing done effects the consecration" (Worship 42), resonates with the postmodern theology of Radical Orthodoxy. Frederick Christian Bauerschmidt, for example, states that "it is through the Church's visible ritual action with bread and wine that the church herself is produced as body of Christ: the agent is produced as it 'exteriorizes' itself in action"; authenticity is an effect of its performance. And Graham Ward discusses not only "the performance of Jesus" but also "the way that performance has been scripted, reperformed and ventriloquized by the community he brought to birth."[34]

Sayers symbolizes the interrelationship of performance and Christian community in *Just Vengeance* when the townspeople nurturing the skeptical airman put on a play for him. Reenacting upon the stage of their world the great drama of the Passion, they perform a series of Bible stories—from the killing of Abel to the killing of Christ—which brings the airman back to faith. Illustrating why church dogma cries out through the stones of the city, the performance draws the airman toward solidarity with the city, leading him to affirm the words of the creed. Because the dogma is the drama, Christianity itself becomes a writing performance.

Sayers's emphatic commitment to the performative nature of church dogma was experientially reinforced by her own participation in drama. In her presentation for the conference at Malvern, written not long before her talk to North London Presbyterian Fellowship of Youth, she parallels the "real community of feeling and interest" between actors with that which should, but usually doesn't, mark the Church: "I recognize in the theatre all the stigmata of a real and living church."[35] Some of these stigmata she had illustrated two years earlier in a sonnet appended to the published version of *The Devil to Pay*. Entitled "To the Interpreter HARCOURT WILLIAMS," the poem honors the man who acted Faustus in this play, as well as William of Sens in *The Zeal of Thy House*. Sayers begins the octave with images of interdependence—"Sound without ear is but an airy stirring / Light without eyes, but an obscure vibration"—and ends comparing these images to drama: "So is the play, save by the actor's making, / No play, but dull, deaf, senseless ink and paper." As Sayers well knew, a play can be created only through the interdependence of heteroglossic positions, a dialogic performance wherein writer, director, actor, scene designer, and costume-maker listen to and learn from each other; for drama to achieve its purposes, the writing must be communally performed. The same holds true when "the dogma is the drama"—as we see happening in *The Emperor Constantine*. In the play, Sayers shows not only the communal writing of dogma during the Council of Nicaea scenes, but also the communal performance of Constantine's faith. The Emperor comes to Christ not through the inner-light of an autonomous conversion experience, but through his involvement with believers throughout his life, from his mother in England to Athanasius at Nicaea, who also guided the communal development of the Nicene dogma.

Dogma, like drama, however, becomes rote when members of the community reiterate their scripts to the point of unthinking mouthing of words and unreflective miming of familiar gestures. Group solidarity starts to solidify dogma into dogmatism, such that protecting the script becomes an end in itself. This is why Rorty posits a need within a community for "strong poets" who can foreground the contingency of the community's vocabulary—to "shake off the dusty phrases" as Sayers puts it (*Ltrs* 2. 354)—providing new metaphors when the old ones become flat, stale, and unprofitable. This, however, leaves Rorty in the midst of paradox. The neopragmatism he advocates, as John McGowan notes, "remains caught between a theoretical vision of the inevitable placement of subjects within historical and social conditions and the notion that freedom resides in breaking through the limits of those conditions."[36]

James Brabazon finds Sayers in the midst of a similar paradox—partly due to Kathleen Nott's *The Emperor's Clothes.* While Sayers disdained Nott's attacks, John Wren-Lewis, the honorary secretary of the Society of St. Anne, believed that the book should be given more consideration in light of St. Anne's goal: to encourage intelligent dialogue between Christians and skeptics. A correspondence developed between Wren-Lewis and Sayers about how best to communicate with the intelligent nonbeliever. At one point Sayers states that she has nothing to offer but dogma, later admitting, however, that "the *history* of the great dogmatic statements" is one of "successive revisions" (*Ltrs* 4. 140, 166). Brabazon comments on this admission, reading it as a fatal flaw contributing to the tragedy of Sayers's life:

> [S]he finally had to make an admission that seems to me fatal to her argument—she had to acknowledge that the Creeds are not in fact statements of ultimate and unalterable truth, but are the result of revisions, based on experience, of earlier drafts; and [according to Sayers] "there is no reason why there should not be other such revisions, except that the schism between East and West has deprived us of the means to call a General Synod."[37]

As in McGowan's criticism of Rorty, Brabazon implies that Sayers remains caught between a theoretical vision of the inevitable placement of subjects within historical and social conditions and the notion that the truth of Christianity breaks through the limits of those conditions. Unlike Rorty, however, Sayers can situate the paradox of her belief in a communitarian construct that celebrates the performance of paradox.

In *Just Vengeance*, Sayers presents the paradoxical nature of the Christian faith in all its complexity, perhaps explaining why she regarded the play as "the best thing I've done."[38] She reiterates a paradox throughout the play: all humans are simultaneously Abel and Cain, because all have the potential for both good and evil, and each person has been both victim and victimizer. As the Recorder notes,

> Do not you all
> Suffer with Abel and destroy with Cain,
> Each one at once the victim and the avenger
> Till Cain is Abel, being condemned for Abel,
> And Abel Cain, in the condemning of Cain? (314)

Sayers relates this paradoxical nature of justice, what she later describes as "the injustice of justice" (*Just* 340), to the current war against the Nazis. As the airman who was killed in that war recounts:

> We drop a bomb
> And condemn a thousand people to sudden death,
> The guiltless along with the guilty. Or we refuse
> To drop a bomb, and condemn a thousand people
> To a lingering death in a concentration camp
> As surely as if we had set our hands to the warrant. (288)

The ethical complexity shown here makes Rorty's neopragmatic ethic seem anemic as he valorizes "liberal ironists" who, among their various "ungroundable desires," include the "hope that suffering will be diminished, that the humiliation of human beings by other human beings may cease" because "cruelty is the worst thing we do" (*CIS* xv). Sayers, I think, would respond by asking "Whose definition of cruelty are you following? The Nazis's or the Jews's? Isn't cruelty itself culturally situated and historically contingent?" Indeed, she has her Recorder in *Just Vengenance* say that "cruel injustice" is often answered "with no less cruel / Justice," contextualizing this comment with the paradox of the current war: bombing Germany leads to the deaths of innocent civilians, hence is "cruel"; failure to bomb Germany will allow the Nazis to continue exterminating the Jews, hence is "cruel." The Recorder reinforces the enigma of justice with a challenge to the city:

> What will you do, citizens, what will you do?
> Since you cannot put down injustice without the Law,
> And the Law is of sin, and turns to sin in your hands,
> Because each one of you is at once Abel and Cain? (341)

In light of this paradox, Rorty's stand against "cruelty" seems like a metaphysical moment in his argument against metaphysics.[39] As Sayers notes in a 1941 broadcast talk, "even the most anti-religious of our thinkers" believe that "fully enlightened humanity" are not "cruel." And then she asks "But why? What makes us so sure about this?" She answers by asserting,

> [A]t the very basis of our thought and and behaviour there lies a pair of assumptions which are wholly religious—which reason cannot prove and for which science can offer no evidence. We assume that both our conception of the good and our human reason are really valid.
>
> . . . .
>
> The assumptions we take for granted about right and reason, and which seem to us self-evident, are not self-evident at all. After nineteen and a half centuries of Christianity, they have become a mental habit with us; but the evidence for them is the evidence for Christianity, and if we reject the one we automatically reject the other.[40]

According to Sayers's paradigm, Rorty, the avowed atheist, is fundamentally religious; nurtured in Western culture, he has unwittingly grounded his repudiation of cruelty in a Christian metaphysic. Indeed, while Rorty feels that Christian belief is no longer pragmatic, he acknowledges that followers of Christ engendered "great moral and intellectual advances of European history" (*CIS* 48). Sayers, had she been given the chance, might have respond to Rorty with a question: "If you're going to establish the paradoxical metaphysic of a groundless ground, why not embrace the paradox of a God who redeems all cruelty through love?" This, in fact, is her point in *Just Vengeance*. In the play within the play, she has Jesus tell Pilate, "there is no justice in the Gospel, / There's only love," having him reinforce this point after his resurrection: "Instead of your justice, you shall have charity" (327, 350).

For Sayers, the paradox of the Gospel reflects the paradox of language: words are human constructions without which we cannot begin to know the ultimate Word who transcends all words. Hence, in *Just Vengeance* she has the "Persona" of God describe the Godhead with, "I the image of the Unimaginable" (318), telling the Virgin Mary,

> Show to them who were made in the image of God
> The image of the Image of the Unimaginable
> From the place where the Image and the Unimaged are one. (323)

The "image of the Image of the Unimaginable" seems to echo Plato, wherein perceived realities are merely shadows imaged on a cave wall in comparison to the Ideal Form that generates them, and the language that attempts to imaginatively capture those shadowy images is even further removed from the Ideal. Plato therefore banned poets and dramatists from his projected utopian Republic. For Sayers, however, images are affirmed. As the Recorder tells the dead Airman in *Just Vengeance*,

> Son, you have come to the place of the images,
> As all men come, whose eyes are not shut fast
> Against redemption, drawn to the moment of glory
> By that god-bearing image, whatever it was,
> That carried the glory for them;....
> Some indirectly—this in a woman's eyes,
> That in a friend's hand or a poet's voice
> Knowing the eternal moment—and you, in the city. (294)

Rather than Plato, we see here the influence of Dante, whose *Divine Comedy* Sayers was reading for the first time as she was conceptualizing the play and which she was translating as she was writing the play's words. Not only is the title, *Just Vengeance*, taken from a line in the *Paradiso*, but Sayers borrows wholesale Dante's "Affirmation of Images"—the concept that humans gain glimmers of the divine not only through the creations of God, but also through those of humanity.[41]

The concept of "creation" distinguishes for Sayers a Christian from a Platonic theory of art. In "Towards A Christian Aesthetic," a lecture she presented the same year she was beginning to read Dante and write *Just Vengeance* (1944), Sayers states that "Plato's conception of art is influenced by his theology—the visible world imitating, copying, reflecting a world of eternal changeless forms already existent elsewhere," whereas the "idea of art as creation"—as something new—is "the one important contribution that Christianity has made to esthetics." Rather than *imitate* Ideal form, artists, Sayers asserts, *image* forth new creations: "There is something that is, in the deepest sense of the words, unimaginable, known to itself (and still more, to us) only by the image in which it expresses itself through creation" (*Whimsical* 79, 83, 84). However, while images bear the divine glory, God is still "Unimaginable," transcending the minds of all makers. And even though Jesus is the Image of the Unimaginable, all images of the Image of the Unimaginable are human creations tainted by the fall, creations that must be "handed over" in an acknowledgment of their inadequacy.

In 1940, the year Sayers was commissioned to create *The Man Born to Be King*, Walter Benjamin wrote, "In every era the attempt must be made anew to wrest tradition away from a conformism that is about to overpower it."[42] Preserving tradition for its own sake implies that a human construction—like the Tower of Babel—is on the same level as God. In his 1937 work *Beyond Tragedy*, from which Sayers quotes several times in her letters, Reinhold Niebuhr includes an essay entitled "The Tower of Babel," which Sayers surely read, given her intrigue with architectural form. The Tower, for Niebuhr, represents how a society sanctifies its own approach to God, as though its historically contingent images contain truth that transcends the ground upon which the images are situated:

> Religion, declares the modern man, is consciousness of our highest social values. Nothing could be further from the truth. True religion is a profound uneasiness about our highest social values. Its uneasiness springs from the knowledge that the God whom it worships transcends the limits of finite man, while this same man is constantly tempted to forget the finiteness of his cultures and civilisation and to pretend a finality for them which they do not have. Every civilisation and every culture is thus a Tower of Babel. (28)

According to the story in Genesis 11, God repudiates such presumption—especially, perhaps, the presumption that human reason can logically determine that no God exists.

Ironically, Kathleen Nott uses the Tower of Babel story to argue against Sayers's defense of plural truths: "the truth about the phenomena and the truth which sustains the phenomena"—in other words, the truth of science and the truth of faith. She counters Sayers by asserting,

> [I]f truth, like Keats's "Beauty," is entirely in the eye of the beholder, anything goes. And there is not only infinite proliferation of entities but there is no way of discriminating their realities: one can go on discussing them but only in the "domain" of Babel.[43]

In *The Emperor's Clothes*, Nott sought to build a Tower of logical reason in defiance of Babel, unaware that her modernist tongue was only one among many Babels of belief. In contrast, postmodernists have reminded us that "the emperor's modernist new clothes are, shall we say, 'culturally constructed' and thereby not universal or absolute sources of truth."[44]

Like postmodern theorists, Sayers acknowledges the many tongues humanity has been given to image forth the Image of the Unimaginable. She therefore develops new glosses for old beliefs, handing over traditional writing performances as she ignites the dross of language with a fiery new tongue. We might therefore describe her with the same words St. Michael spoke to William of Sens in *The Zeal of Thy House*: "For you the task, for you the tongues of fire" (98).

Chapter 6

Begin Here: For the End(s) of the Performance

Life is an interminable serial story, packed with exciting episodes. We cannot read up all the back numbers in full, nor can we wait to see the end before we begin to read it.

—Sayers, *Begin Here*

Few people would think to end a book on Dorothy L. Sayers with an analysis of *Begin Here*, Sayers's "memorandum of what some of us in Britain were thinking and feeling in the Winter of 1939."[1] Usually dismissed by scholars due to its hasty inscription and resulting errors, *Begin Here* nevertheless alludes to a fundamental problem addressed by theorists of our own time: the problem with assessing origins, both of life and of art. If, as suggested by this chapter's epigraph, "life is an interminable serial story," who authorizes the writing performance and how much power does the hero have to change the plot? Since the "back numbers" of life are not fully accessible, how might we build upon or challenge what "History" has taught us about ourselves so that we can perform responsibly? (*Begin* 19). Upon what authority does any human act? When and where does the self begin? The difficulty of answering these questions may explain why *Begin Here* "ranges widely, and somewhat superficially, through history, economics, theology, and philosophy."[2] Sayers is wrestling with the legitimacy of human agency as she struggles to assess how much control one has to direct the "exciting episodes" in one's own life, let alone to change the world at large. As we shall see, her questions and concerns parallel interrogations and analyses of "discourse" generated by preeminent shapers of postmodern thought: Jean-Francois Lyotard, Michel Foucault, Roland Barthes, Jacques Derrida, Judith Butler. "Discourse," for these theorists, alludes to the performativity of language, wherein knowledge and behavior are formulated according to the citation and reiteration of cultural scripts. Thus, what we assume to be "true" is actually the effect of the unrecognized and unquestioned assumptions embedded in language. Human behavior becomes for them a type of writing performance. Sayers, however, while recognizing the performativity of language, escaped "the prison house of language," as Fredric Jameson famously described it, due to her Anglo-Catholic assumptions. She believed that humans maintain a freedom of choice that gives performances, written or otherwise, meaning and value.

The performativity of language might be illustrated by way of Sayers's fourth mystery novel, *The Unpleasantness at the Bellona Club* (1928). The story opens with the exclamation, "What in the world, Wimsey, are you doing in this Morgue?" thus signaling a fundamental convention of detective fiction. After all, the presence of a dead body is necessary to generate the process of detection and thus the narrative itself. However, in *Bellona Club*, Wimsey's follow-up retort—"Oh, I wouldn't call it that... Funeral Parlor at the very least"—reveals that the word "morgue" was merely a metaphor used by George Fentiman to describe the lugubrious atmosphere of the Club. Fentiman goes on to call the old gentlemen snoozing in their chairs "corpses," farcically quoting a line from *Punch*: "Waiter, take away Lord Whatsisname, he's been dead two days" (1). By the end of the chapter, however, we discover that one of the old gentlemen is, indeed, dead, and that the corpse is that of Fentiman's grandfather. The metaphoric word-play that begins the chapter thus becomes hypostatized as word is instantiated in flesh.

The stiffened corpse discovered in a wing-chair—literally turning the Bellona Club into a type of morgue—could operate as an analogy for certain poststructuralist views of the body, wherein language is the "very condition under which materiality may be said to appear."[3] After having metaphorically described his grandfather as a corpse, George Fentiman soon discovers the body is actually dead. He can talk about the material reality, however, only by performing language already written, quoting once again from *Punch* to hysterically assert "He's been dead two days! So are you! So am I! We're all dead and we never noticed it!" (5). Fentiman's word "dead" vacillates between the literal and the figurative—an actual dead body surrounded by spiritually dead observers—drawing attention to the contingency of its meaning. The word performs differently depending how it is staged.

Forty years after this *Unpleasantness*, Roland Barthes proclaimed another unpleasant death: "the Death of Author." Barthes thus signalled his sense that authors—indeed, all humans—don't autonomously generate thoughts that they put into words. Instead, the words of a culture precede and hence constitute thought—just as words from *Punch* constituted the thoughts of Fentiman. Since, for Barthes, all writing performances are previously scripted, a reader does not access the "original" ideas of an author, but instead interprets the multiple linguistic strands threaded through her mind that are knotted together in the text. Therefore, authors do not perform thought as they write; instead, writing performs through them. As Barthes puts it, "to write is, through a prerequisite impersonality . . . , to reach that point where only language acts, 'performs,' and not 'me.'"[4] The impersonal priority of the word, then, leads to the "death" of authority as it has traditionally been known: authors do not authorize their language. Words cause meaning beyond an author's control, as when Fentiman's word "morgue" anticipates a meaning that exceeds his intention.

The "Death of the Author" did not worry Sayers as she wrote *Begin Here*; the very concept was far removed from the modernist assumptions molding thought in 1939. However, Sayers shared with Barthes an interest in the way language authorizes competing definitions of what it means to be human. Especially concerned with "the tendency to regard Economics itself as an absolute standard of value," she describes how language reflects and controls our economically motivated perceptions:

> We shall hardly realize how deep this tendency has rooted itself in our minds till we stop to examine the language we habitually use and see how often, for instance, we use the word "work" as though it meant the same thing as "gainful employment," how instinctively we talk of "a big film," meaning no more than that it cost a great deal of money to produce. (*Begin* 101)

Sayers here recognizes that discourse—"the language we habitually use"—performs our perceptions about "work" in demeaning ways.

Economics is only one stage upon which "truth" has been performed. In *Begin Here* Sayers describes the various theaters of representation that have scripted Western culture since the "collapse of spiritual standards as an absolute authority in world affairs." After Humanists established "Man" as "an absolute value in himself, apart from any outside standard," Reason became the "supreme object of human devotion," such that poetry, architecture, and even nature came to "express the dominance of rationality in man's notion of himself." This celebration of rationality was embedded in the notion of "progress"; however, while the so-called Age of Reason valorized "Progress" employing a mechanistic script, the century of Darwin situated "progress" in an "organic theory of the universe," where "survival of the fittest" was not only biological but also economic, with "wealth acknowledged as an absolute authority in itself." With each stage, a new authority becomes absolute, each one performing under "the sacred name of progress" (*Begin* 63–67).

Sayers's word "sacred" signals why she glibly covers five centuries of European thought in five pages. She does not chronicle history so much as illuminate the displacement of transcendent authority onto human constructs, like the one defining the Aryan self in 1939: "Nazism has gripped the German people with all the power of a militant religion: it *is* a religion—the ancient, mystical, persecuting religion of blood and race" (*Begin* 128). As with the word "myth" discussed in chapter 5 of *Writing Performances*, Sayers uses the word "religion" to denote a "metanarrative," the term Jean-Francois Lyotard employs in *The Postmodern Condition* to describe a totalizing account for "reality," what Sayers calls a "complete explanation of the universe" (*Begin* 47). The metanarrative that Lyotard finds especially problematic is that of the "Enlightenment," wherein the exercise of reason, especially scientific reasoning, provided objective "truth" leading to the betterment of society. Lyotard considers all metanarratives to be "terroristic," for they marginalize and often dehumanize people who interpret reality according to a different paradigm. Though his word "terroristic" seems somewhat apocalyptic to describe the metanarrative of Enlightenment Humanism, the term is entirely appropriate for the metanarrative of Nazism that concerned Sayers: "Nationalism as an absolute principle" (*Begin* 88). The writings of National Socialism staged the "human" in appalling performances.

Sayers reiterates the words "authority" and "absolute" in *Begin Here* as she describes the metanarratives that authorize the performance of "truth" in different eras. She thus makes her point that, "with the abandonment of an absolute Authority outside history, the seat of absolute authority within history tends to become identified with the seat of effective power" (*Begin* 60). As Lyotard was to note forty years later, "knowledge and power are simply two sides of the same question: who decides what knowledge is, and who knows what needs to be decided?" (8–9). Around the same time, Michel Foucault was also speaking out against the "tyranny" of "globalizing" vocabularies. A profound influence on postmodern thought, Foucault, like Lyotard, questioned the authority of Enlightenment and Humanist "truths" about human "progress."[5]

Regarding the idea of "progress" as a problematic human construction, Foucault, I believe, would have condoned Sayers's view that "protests of Religion, Humanism and Reason against the effect of Capitalism were greatly hampered by the fact that they themselves had capitulated to the biological theory of 'progress' " (*Begin* 69). Neither Sayers nor Foucault, though both sounding quite Marxian at times,[6] see Communism as the answer to the economic metanarrative of Capitalism; for Marxism has "capitulated" to the absolutism of "progress" as well. Sayers writes,

> It has been claimed for the Marxian theory of society that it is "entirely free from all dogma"; but that is nonsense. All theories of society are founded on dogma—what causes a dogma to go unrecognized is its unquestioned acceptance; it is not recognized, that is, as dogma when it has sunk so deeply into the structure of our thought that we take it for granted. (*Begin* 78)

Sayers's word "dogma," like her word "religion," is a bit problematic, since it denotes willed determinations of meaning as church councils formulate creeds based on the "opinion" of a body of believers. Sayers, however, qualifies her definition of "dogma" through reference to "unrecognized" and "unquestioned" acceptance of concepts that "structure our thought," implying that ideas we think to be naturally self-evident or commonsensical may actually be social constructions embedded in discourse.[7] As she notes in her introduction to Dante's *Purgatory*, "Those preferences which appear to us self-justified out of all argument belong to a period attitude, no less (if no more) than the corresponding preferences of fourteenth-century people" (45). Hence, when a contemporary of Sayers like I. A. Richards argues that "most critical dogmas of the past are either false or nonsense," he merely performs the "period attitude" of modernism, which assumes that enlightened humans have progressed beyond a need for religion.[8] The postmodern Foucault, however, like Sayers before him, regarded this idea of "progress" as simply another dogma.

In his early work, Foucault uses the word "episteme" to refer to the constellation of unacknowledged dogmas that constitute the perceptions and practices of an era. Like Sayers, he sees the Enlightenment episteme as shaping the history of science into a teleological structure, with progress achieved through rational exploration. However, since "rationality" is itself an artificial construct, the discoveries of science are situated in structures of belief, not in individuals. Sayers makes a similar point in *Begin Here*:

> Reason, however, before it could proclaim the "perfectibility of man," was forced to make an act of faith, since—if one left God out of the picture—there was no evidence whatever that evolution was bound to end in *moral* perfection, or in perfection of any sort, for the human animal. Other animal types had remained static, or had degenerated and been superseded, and why not *homo sapiens*? The act of faith was, however, made; and Reason professed belief in its own perfectibility, though at the cost of denying its own authority. (66)

Like Lyotard and Foucault, Sayers questions what authority grounds the truth-claims of reason. For Foucault, there is no such thing as "disinterested truth," for knowledge is always "regularized and constrained" by "power."[9]

Some people have misinterpreted Foucault's use of "power" as comparable to Marxian "ideology" wherein the bourgeoisie reinforces self-serving dogmas that keep the proletariat subdued. Foucault, however, regarded Marxism as naive not only in its assumption of progress, but also in its belief that there is an essential nature in humans waiting to be liberated from restrictive economic systems. For Foucault, power originates neither in human subjects nor in social institutions but is diffused throughout society. Indeterminate in character, with no origin, Foucauldian "power" nevertheless generates human thought and behavior. As Judith Butler puts it, "There is no power that acts, but only a reiterated acting that is power in its persistence and instability."[10] Foucault therefore undermines human agency through his theory that one's "identity," better called one's "subjectivity," is merely the "subjection" of the self to discourse: the performance of unrecognized dogmas that not only normalize behavior but also shape any revolt again them. Since it is therefore impossible to be free of the discourses that script human performance, he declares the "death of the Subject."[11]

Though suspicious of language, Sayers would have had no patience with the "death of the subject." Indeed, her concern in *Begin Here* was how to take action in light of the *literal* deaths of subjects at the hands of the Nazis. Sayers therefore contemplated human agency not in contrast to the constructed nature of subjectivity, but in order to argue that "peace had to be actively constructed." She exhorts her readers to consider life as "perpetual activity," with "peace conceived as the energetic balance of liberty and equality, mercy and justice, truth and charity" (*Begin* 136, 138). But how can one take a stand for mercy, justice, truth, and charity if these abstractions are merely "actively constructed" performances of discourse that reflect the "period attitude" of a particular society at a certain point in history? How can one condemn Hitler's actions as "evil" if all behavior is merely a writing performance?

Sayers's concerns for political engagement and its relationship to the "dogmas" that define "the Whole Man" anticipate the perspectives of many late twentieth-century theorists who took a stand against Foucault's "death of the Subject." For them, as Geraldine Harris notes, "the focus in both postmodernism and poststructuralism on notions of discourse, linguistic theory and signifying systems is seen as producing a movement away from any active engagement with the material world." A concern with "agency"—the ability to choose work that subverts the problematic performances of cultural dogmas—drives those critics of radical poststructuralism who want to hold onto distinctions "between right and wrong, good and bad, fake and copy."[12] They believe that to assume the inescapability of discourse is to submit to the inertia of the status quo—what Sayers calls "sloth" in *Begin Here*. In order to remediate injustice, they would argue, one must situate agency outside of the dogmas that construct behavior. As Elizabeth Abel notes, their concern has "fostered a turn to texts that reassert the authority of experience, that reinstate political agency, and that rearticulate the body and its passions."[13] Many of these critics would agree with Sayers's statement that "Knowledge does not become thought till we have made it part of our lives by relating it to our experience and acting upon it" (*Begin* 11–12). But what can give authority to "our experience"? Upon what stage does the performance of life generate not just a display of performativity, but meaning and value? Sayers would answer (had she access to postmodern diction) that a ground must be established outside of the performativity of language—a prediscursive site—upon which to situate a concept of the self that is able to challenge the unrecognized dogmas of society. This does, indeed, seem to be a goal of her wartime message: to *begin* to perform responsibly, one must *begin* with an absolute that transcends all culturally constructed writing performances.

Many postmodern advocates of agency begin with the body, which "is still there, beyond theory."[14] After all, the pain of abuse is experienced by the body even when hegemonic discourses assert that African slaves are not fully human, that women do not have the same intellectual capabilities as men, that homosexuals should be reviled. The body can "know" injustice even when the discursive practices of a society might legitimize lynchings, limited suffrage, or abusive hate speech. Hope in the body as a pre-linguistic origin of signification becomes what Kathryn Bond Stockton calls a "real-bodies mysticism": a "belief in something real that escapes and exceeds human sign systems."[15]

Sayers, of course, believed in something real that escapes and transcends human sign systems, situating human agency not in the prediscursive site of the body but in the sight of a prediscursive God—who, according to Christian dogma, took on the form of a human body. Hence, her criticism of the arbitrary "absolutism" of cultural "dogmas" arises from a fundamental belief in an eternal Absolute outside of history which is the origin of human meaning and hence the only means to provide authority for human agency:

> [I]f we deprive ourselves of the eternal Absolute, we shall inevitably make an absolute of some temporal thing or other—be it Liberty, Equality or Progress . . . Race or Reason, or even Unreason, or the Perpetual Flux of Relativity. (*Begin* 127–28)

Foucault, like many poststructuralists, would respond that her very concept of an "eternal Absolute" is performative: an effect of discourse. He therefore ends his most famous work of the 1960s, *The Archaeology of Knowledge*, with a philosophic extension of Nietzsche: "you may have killed God beneath the weight of all that you have said; but don't imagine that, with all that you are saying, you will make a man that will live longer then he" (211). Ironically, I think Sayers would agree with this statement, for it implies that, without an eternal Absolute, humans are merely performing puppets, tied to strings of discourse. Sayers most certainly believed that, without God, we face the "death of the Subject." In *Begin Here* she asks "Are we helpless puppets in the hands of irrational forces . . . ceaselessly speeding hither and thither without direction?" (48–49). For her, agency must be established in something that exceeds writing performances.

People unsympathetic with Sayers's theistic assumptions might protest that this implied parallel between "the body" and "God" is unfair. After all, the body is an empirically verifiable material reality whereas the existence of God is not at all verifiable—unless one appeals to the High Church emphasis on dogmas sustained by an interpretive community or to the Low Church emphasis on narratives of personal relationship with Christ. They would note that, while the body is visible and touchable, personal "experience" of an unseen reality can be interpreted either psychologically or poststructurally: as the performativity of faith vocabularies that construct rather than refer to the phenomena they describe. And the Bible that guides both High and Low church practices can be similarly dismissed: either as a narrative intentionally inscribed to cover over humanity's sense of lack—a lack of meaning and power—or as the record of discursively constituted fictions. There does not seem much that a Christian like Sayers might say in response, since faith, according to the author of the New Testament letter to the Hebrews, is, indeed, "the assurance of things hoped for, the conviction of things not seen" (RSV 11:1). However, postmodern theorists assert that conviction situated in any metanarrative is a thing hoped for, arguing that all knowledge, whether of the scientist or the Sunday School teacher, is embedded in the "dogma" of a discursively generated tradition.

Judith Butler goes so far as to argue that even assurance about a prediscursive body is simply a thing hoped for, that convictions about material realities are merely performative. This is not to deny that bodies matter, but to say that even the way we see and touch the body is an effect of discourse. For example, we see a corpulent body differently depending on when and where we live. In one culture it is a material sign of success; signaling the abundance of wealth, it becomes a signifier of desirability. In another culture corpulence implies lassitude and sloth, signaling what must be avoided for the body to be a sign of success. Thus, discursive practices (like gormandizing or dieting) depend upon and perpetuate the performativity of the word "stout": the material effects of the body do not lie outside discourse.

Butler concludes, then, that establishing the prediscursive body as an origin for the self is an illegitimate metaphysical move, arguing instead for the inescapable performativity of language. However, she also dislikes the way some theorists advocate what she calls "linguistic monism," saying that radical constructivism "has taken the place of a godlike agency which not only causes but composes everything which is its object; it is the divine performative, bringing into being and exhaustively constituting that which it names."[16] Sayers, of course, would agree that this is problematic. *Begin Here*, as we have seen, recounts numerous examples of discursive substitutes for the "eternal Absolute." Butler, however, would deny the possibility of any "Absolute," using the words "godlike" and "divine" as pejorative adjectives tied to fictions that must be dispelled. She most surely would state that, as soon as one posits a God situated outside of language, such positing is itself a performance of language, so that discourse is still the constitutive condition by which God exists.[17]

Denying any essential meaning outside the performativity of language, Butler must reconceptualize the authority for agency in oxymoronic terms. She speaks of "authority's groundless ground" (108) by which a self can act: "The force of repetition in language may be the paradoxical condition by which a certain agency—not linked to a fiction of the ego as master of circumstance—is derived from the *impossibility* of choice" (124). Butler's *agency*, a word that implies the choice to act, is allowed no choice; any ground for action is groundless. I would argue that Butler's paradoxes are no less daunting than the paradox which gave authority to Sayers's sense of agency: Creator God took on created flesh, affirming simultaneously the Body and God as origins of meaning. This, of course, must be believed on faith, but even Butler admits feminists often take a "faith" position in order to legitimize agency: "feminists need to operate on an operational essentialism, a false ontology of women as universal in order to advance a feminist political programme."[18] Once again, she is reduced to paradox: though the essentialism is false, it must be regarded as true.

A similar contradiction inscribes many postmodern discussions of agency. Antifoundationalists who want to avoid ethical relativism inevitably establish some kind of ground upon which to build their ethics, whether love, justice, intuition, or the body. Sayers, like many people of faith, is simply more explicit about her ethical ground; it is situated upon God as the source of existence: a pre-discursive site of origin. While many contemporary theorists might eschew such grounding as a metaphysical crutch, they are left with the paradox outlined by Bradley Butterfield: "rejecting all metaphysical first principles, one is left no ground from which to launch a critique of metaphysics."[19] This may explain why feminist scholars Julia Kristeva and Luce Irigaray increasingly focused upon God in their concern for female agency, and African American critics Cornel West and bell hooks claimed God as the ground upon which they situate their racial politics.[20] Likewise for Sayers, belief in the Eternal Absolute, the abjected "Other" of the modernist discourse of her own time and of many postmodern discourses today, is the only way to ground an ethic of love toward the "other."

In *The Mind of the Maker* Sayers repeatedly appeals to "love" as the constitutive attitude of the Creator to Creation: "a work of creation is a work of love" (129). Furthermore, "an inner voice reminds us that the Christian God is Love, and that love and resignation can find no common ground to stand on" (136). For "resignation" we could substitute the words "inertia" and "sloth," those attitudes Sayers denounces in *Begin Here*. Love is the source of agency, and God is the source of love, all of which is intuitively perceived through an "inner voice." Therefore,

> When there is a genuine conflict of opinion, it is necessary to go behind the moral code and appeal to the natural law—to prove, that is, at the bar of experience, that St. Francis does in fact enjoy a freer truth to essential human nature than Caligula, and that a society of Caligulas is more likely to end in catastrophe than a society of Franciscans. (*Mind* 10)

Though it has long been anathema among poststructuralists to speak of "essential human nature," most people in most societies would probably prefer living with St. Francis, an icon of other-oriented agency, rather than with Caligula, an index of self-interest. Sayers thus justifies the "natural law" of love "at the bar of experience," to which she refers repeatedly in *The Mind of the Maker*.[21]

Sayers's appeal to intuition and experience might be considered naive by some twenty-first century critics; however, it is duplicated in many postmodern interrogations of ethical agency. In *Situating the Self: Gender, Community and Postmodernism in Contemporary Ethics*, Seyla Benhabib uses both "experience" and "intuitive ideas" to justify what she calls "interactive universalism" wherein "we recognize the right of all beings capable of speech and action to be participants in the moral conversation" (24). Sayers, I believe, would wonder what legitimizes Benhabib's "experience" and "intuitive ideas"; either they are constituted by the performative "dogmas" of a community, and hence are not universal, or else they are tied to something that transcends our human situatedness. For Sayers, those who hold to a purely naturalistic explanation of existence, an existence Darwin quite convincingly situated in natural selection, are hard-pressed to explain why an intuition for justice should ever extend beyond the selfish gene of an individual species or tribe.[22]

In contrast, Sayers grounds her thought in an "eternal Absolute" that transcends the limitations of discourse.[23] Because "all language about everything is analogical," truth exceeds language: "we think in a series of metaphors. We can explain nothing in terms of itself, but only in terms of other things" (*Mind* 23). Sayers made this same point in a lecture to the Oxford University Socratic Society in 1954, outlining historical analogies for the doctrine of the Atonement to illustrate that even theological language changes with the discursive practices of societies. This only gets "perilous" when "image and reality are too literally equated" by people who want to absolutize linguistic meanings that serve their own interests, a perspective she parodies with the words, "Hold on and squeeze hard, and, after it has changed shape a thousand times, language will tell you what you wish to know" (*Search* 283, 286). Signifiers are the result of human performances, unable to make present the signified reality toward which they gesture; however, they are all we have with which to create new perceptions. As Derrida notes, "we cannot do without the concept of the sign."[24] Sayers uses a similar argument in *The Mind of the Maker* in order to legitimize vocabularies about God:

> Skeptics frequently complain that man has made God in his own image; they should in reason go further (as many of them do) and acknowledge that man has made all existence in his own image. If the tendency to anthropomorphism is a good reason for refusing to think about God, it is an equally good reason for refusing to think about light, or oysters, or battleships. (23)

For Sayers, if we're going to think at all, we might as well think about an eternal Absolute that provides an origin, and hence meaning, to the subject that performs ethically. This helps explain why Sayers was such an energetic defender of Christendom's dogmas; though constructed out of language, Christian dogmas have lasted thousands of years, providing core writings which have performed Christianity through the ages.

Like her perspectives on Anglo-Catholic worship explored in chapter 5, Sayers's perceptions in *Begin Here* anticipate what Laurence Hemming calls the "postmodern performance" of Radical Orthodoxy. As noted by Graham Ward, "Radical Orthodoxy's concern is with unmasking the cultural idols, providing genealogical accounts of the assumptions, politics and hidden metaphysics of specific secular varieties of knowledge." Similar to Sayers, then, proponents of Radical Orthodoxy defy the constructs of Enlightenment modernism, aware "that authorities are ephemeral and open to challenge, that facts and the brutally given are plastic and malleable to interpretation, that representation (both institutional and discursive) involves distortion."[25] Nevertheless, Radical Orthodoxy, like Sayers, endorses belief in traditional Christian verities, especially those expressed by Augustine and Aquinas, over agnosticism. As Sayers notes of the *Man Born to Be King* scandal,

> This confirms my general impression that if there is anything at all that was *not* dealt with by Augustine, it was dealt with by Aquinas, and that if one could give one's self up to read the whole of the works of these two saints one would never need to read anything else. (*Ltrs* 2. 365)

Radical Orthodoxy's "skill in elegantly re-performing Aquinas" often parallels the skill of Sayers's writing performances. Laurence Hemming summarizes an assertion of Aquinas in *de Veritate* with words that sound straight out of *The Mind of the Maker*: "our minds are really analogous to God . . . in the faculty of creativity itself." John Milbank, one of the founders of Radical Orthodoxy, wants to push beyond Aquinas's analogy to assert the "fundamentally poetic being" of life itself: a performed poetry of faith. "A doctrine of full moral responsibility," as he states in *The Word Made Strange*, requires the "*faith* that if we attend to God, he will graciously provide us, out of ourselves, with appropriate good performances." Performing the *Imago Dei* through ritual as well as art, "radically orthodox" Christians create new meanings while simultaneously performing an ancient script, their view of creativity "modelled upon a Trinitarian and incarnational understanding of participation."[26]

"Participation" is also key to Sayers, who celebrated the "Divine joy in creation, which Man should inherit in virtue of his participation in the image of the Godhead" (*Ltrs* 2. 251). This, then, is how Sayers grounds ethical agency. Just as God was, and still is, the agent of Creation, so humans, created in the image of God, exercise agency through creative energy:

> We cannot deal with industrialism or unemployment unless we lift work out of the economic, political and social spheres and consider it also in terms of the work's worth and the love of the work, as being in itself a sacrament and manifestation of man's creative energy. (*Mind* 218)

Work that becomes identified with "gainful employment," rather than with "joy in creation," allows for "wheat and coffee to be burnt and fish to be used for manure while whole populations stand in need of food" (*Creed* 52). Sayers, like the advocates of Radical Orthodoxy, thus displays concern over the abuses of Capitalism:

> The avaricious greed that prompts men to cut down forests for the speedy making of money brings down a judgement of flood and famine, because that sin of avarice in the spiritual sphere runs counter to the physical laws of nature. (*Creed* 49).

By tying agency to the performance of created natural laws—like the dire results of deforestation—Sayers can encourage performances that not only support creation but also perform creatively.

In her commitment to the agency in and of creation, Sayers resists concepts of unsurmountable determinism, whether of biology, psychology, economics, or discourse:

> The final tendency of the modern philosophies...has been to bind man hard and fast in the chains of an iron determinism. The influences of heredity and environment, of glandular make-up and the control exercised by the unconscious, of economic necessity and the mechanics of biological development, have all been invoked to assure man that he is not responsible for his misfortunes and therefore not to be held guilty. (*Creed* 47–48)

She therefore also rejects views of God promulgated by Christian "schools of thought" which assert that everything "is rigidly determined for us" (*Creed* 13). Repeatedly throughout her writings, Sayers defines her faith in contradistinction to the "total-depravity" and "determinism" characteristic of many Calvinists, believing that both constructs limit the sacramental agency of work. While affirming the power of God, she nevertheless questions views that limit the free will of humans: "Why did He not make us mere puppets, incapable of executing anything but His own pattern of perfection?" (*Creed* 13). This perspective, as we have seen, also informs *Begin Here*: while affirming the power of discourse, she questions views that limit humans to "helpless puppets in the hands of irrational forces" (48). Radical determinism, whether of God or of language, leads to a "dreadful conclusion": if one "is not responsible for evil, he cannot alter it" (*Creed* 48).[27]

Mikhail Bakhtin shares Sayers's desire to legitimize agency, so that "ethical freedom ('freedom of the will') is not only freedom from cognitive necessity (causal necessity), but also freedom from aesthetic necessity."[28] As he states in *Toward a Philosophy of the Act*, "to be in life, to be *actually*, is to *act*" (42). He situates the ability to act, the agency behind "Being-as-event," in "the validity of truth...sufficient unto itself, absolute, and eternal" (10), what Sayers calls the Eternal Absolute. Bakhtin writes:

> Outside God, outside the bounds of trust in absolute otherness, self-consciousness and self-utterance are impossible, and they are impossible not because they would be senseless practically, but because trust in God is an immanent constitutive moment of pure self-consciousness and self-expression.[29]

Ethical agency, a philosophy of the act, participates in the relational nature of God, a creator who cares for the transformation, the becoming, of creation. For Bakhtin, then, an ethically performed act (what Sayers calls "the integrity of work") should be motivated and determined by the "ought" of the future, rather than the "is" of present circumstances; it is not situated solely in who I am now, or what culture makes of me, but in what I may become through its performance. As Sayers once put it, "What we make is more important than what we are."[30]

Writing Performances: The Stages of Dorothy L. Sayers has traced the "making" of Dorothy L. Sayers, whose writing performances were staged in different theaters of representation as she grappled with the theatrics of representation itself: of autobiography, of gender, of art, of creativity, of Christianity, and finally of that which exceeds all representation: "the Unimaginable." However, believing that the Image of the Unimaginable walked this earth, Sayers took joy in creating images, in performing the *Imago Dei*. Because she saw "the temporal universe as one of those great serial works of which installments appear from time to time" (*Mind* 59), life itself became, for her, "an interminable serial story" (*Begin* 19). Human life is God's writing performance.

And the story continues after death. Scholars, through their own writing performances, add new chapters to Sayers's work as they read it from their own historical, theological, and philosophical standpoints.

I, as well, begin here.

Encore: A Conclusion

Let each do well what each knows best,
Nothing refuse and nothing shirk,
Since none is master of the rest,
But all are servants of the work—

—Sayers, "The Makers"

Like any good writer of detective fiction, Dorothy L. Sayers placed in her narratives "red herrings" that seem to identify the perpetrator of the crime. When attentive readers recognize that a red herring has led them astray, they change the direction of their thoughts, following a new scent as they seek, along with the detective, the identity of the culprit. In *The Five Red Herrings*, her seventh detective novel, Sayers personifies such identifying characteristics as six characters, all of whom seem capable to have committed the murder of an artist named Campbell. She provides motives that make each appear, for a time, as a likely suspect. By the end, however, the reader recognizes that what often seems to be proof positive of the criminal's identity is actually grounded in the factitiousness of fact.

Writing Performances has sought to demonstrate that identifications of Sayers are also grounded in the factitiousness of fact, as biographers and critics interpret the writing performances of Sayers and those who knew her in different ways. In the history of Sayers scholarship, the woman herself has become five red herrings: the aspiring poet, the writer of best-selling detective fiction, the playwright behind several successful productions, the prophetic literary theorist, and the unorthodox apologist for Christian orthodoxy. But no single one of these characters is responsible for the life of this extraordinary person.

In *Five Red Herrings* the character responsible for the life of the murder victim seeks to cover-up his crime by making Campbell's death seem like an accident. Ferguson, the culprit, paints a picture in Campbell's style, leaving the canvas at the edge of a cliff so that others will conclude Campbell tripped and fell into the gorge below. Several characters try to identify who executed the painting, hoping to detect who executed the painter. Sayers thus uses painting, as she did in *The Documents in the Case* and *Thrones, Dominations*, to problematize identity, forcing us to acknowledge that pictures can never fully capture the real self. One of the novel's five red herrings comments on the painting, "I call that a thoroughly Campbellish piece of work. A poor Campbell, if you like, but full of Campbellisms." Lord Peter responds by paralleling painting with writing performances: "G.K. Chesterton says . . . that most people with a very well-defined style write at times what look like bad parodies of themselves. . . . I expect painters do the same" (219). Because the performance of identity is an effect of the performativity of style, whether strokes of paint on a canvas or ink on paper, the "real" Dorothy L. Sayers varies according to which theater of representation stages her writing performances.

By the last stage of her career, Sayers saw herself parodying another person's style altogether. Like Ferguson painting "a Campbell," Sayers was painting Dante, "who was the friend of painters, and who (as we know) sometimes did a little painting himself, [and] had the painter's eye."[1] Attempting to capture his "colour" in a colloquial English terza rima, Sayers was so immersed in Dante's writing performance that when she wrote *The Just Vengeance* for the stage, she found that "nearly all the argument and most of the best lines are Dante's" (*Ltrs* 3. 205). Dante's language started to perform her, apparent as well in the last decade of her correspondence, which is filled with Dante analysis, quotations, allusions, and metaphors.

At the same time, Sayers recognized that her writing performances changed Dante. Like her "traditioning" of orthodox Christianity, regarded by some as a "betrayal," Sayers's translations performed Dante in a controversial style. As she was to tell C. S. Lewis, "I'm really experimenting in a new way of doing Dante, and it's quite likely that I have fallen into errors of taste here and there" (*Ltrs* 3. 465). Nevertheless, she recognized that tradition, true to its etymology, needs betraying:

> Everybody will say, Who am I to set myself up against six centuries of better scholars? Nobody, indeed. But if the traditional interpretation leads everybody sooner or later into the ditch, it may be . . . that a fresh eye, though—or even because—untrained, might happen to spot the error. (*Ltrs* 3. 516–17)

Sayers's "fresh eye," situated in a different place at a different time than those of earlier scholars, can exercise what Bakhtin calls an "excess of seeing" that "consummates" Dante in new ways—through writings that remediate the biases of the past.

Sayers, however, recognized that even her own writings, dependent upon the cat-like vagaries of language, were not entirely under her control: all writing performs according to the biases of its time. In fact, as Sayers reminds us in *The Documents in the Case*, life itself "is a kind of bias—a lop-sidedness." Like the Dante scholarship that needed a fresh eye, the lopsidedness of life necessitates an excess of seeing—something Sayers recognized when she began writing performances for theater, not only through the plural perspectives she put on stage, but also among the multiple talents putting up each play. This dialogic creativity led her to theorize about the very nature of creative work. Believing that God is plural, the dialogism of a trinity, Sayers argued in *The Mind of the Maker* that human creativity performed the *Imago Dei*.

Sayers's sense of her writing performances might be summarized through a poem written to preface her published version of *A Man Born to Be King*. In "The Makers," she creates a conversation among an architect, a craftsman, and "the stone," all of whom claim responsibility for architectural form. While the architect, like a playwright, develops the "thought," "dream," and "plan" behind the work, the craftsmanship of the carpenter, like that of actors performing on stage, makes the plan visible. As the craftsman tells the architect,

> It is by me the towers grow tall,
>   I lay the course, I shape and hew;
> You make a little inky scrawl,
>   And that is all that you can do. (13–16)

Sayers thus sets up, in the first five stanzas of her poem, an architectural paradigm reminiscent not only of her early poetry, but also of the trinitarian aesthetic that she developed in *The Mind of the Maker*: the Idea of the architect becomes instantiated through the Energy of the craftsman. We assume, then, that the "stone" must represent the Power of the audience, which "can bless or damn" the joint creation of architect and craftsman (22). However, this interpretation seems less viable when the stone alludes to "the cleaving steel, the crushing load / That make a calvary of me":

> And you may carve me with your hand
>   To arch and buttress, roof and wall,
> Until the dream rise up and stand—
>   Serve but the stone, the stone serves all. (31–36)

Addressing both the architect and the craftsman, the stone stands not for the audience, but for language, that fundamental material condition of writing performances.

In fact, Sayers refers to the language of stone in several of her plays, alluding, each time, to a famous Bible verse: "If these should hold their peace, the stones would immediately cry out" (Luke 19:40). In *The Man Born to Be King*, of course, she has Christ perform this verse as part of his triumphal entry into Jerusalem. In *The Zeal of Thy House* it is the stones of Canterbury Cathedral that metaphorically "cry out." For *The Just Vengeance* Sayers has "the stones of the city" cry out.

In "The Makers," the stone cries out about becoming a "calvary." Language, indeed, became a type of calvary in Sayers's experience of *The Man Born to Be King*, with early audiences protesting that she had martyred the Truth, having placed the incidents of Jesus' life upon a hill of colloquialisms. The protesters, as we have seen, wanted Sayers to use only the golden, sacred language of the "Authorized Version" of the Bible. However, as the stone argues in "The Makers," "I am granite and not gold . . . I am marble and not clay" (25–26). Though words can be cloven, crushed, and carved into new forms of expression, they maintain an obdurate power which, like the stoniness of granite and marble, precedes human thought. Or, as the Gospel of St. John puts it, "In the beginning was the Word" (1:1). In orthodox Christianity, God did not create the Son; so also an idea does not mould language, as if it were "gold" or "clay," into a vessel of signification. Language is the "master" of thought. Hence, the stone in "The Makers" protests, "You may not hammer me nor mould— / I am the master of the way" (27–28). Language is the bedrock of human performance, both written and otherwise.

While Sayers gives the architect the first two stanzas of "The Makers" and the craftsman the next three, the stone speaks for eight stanzas, including the epigraph with which this Encore begins. By suggesting that "all are servants of the work," the stone asserts the need for agency—for "work"—which can escape the determinism of language while simultaneously acknowledging its granite and marble inexorability. Sayers's acceptance of this paradox is what distinguishes her as a distinctive mind in an era that regarded Christianity as a feeble performer in the rationalist theater of empirically verifiable truth. By recognizing that human reason is itself constructed from the stones of language, Sayers anticipated postmodern paradigms. However, due to her belief in a pre-discursive Author(ity) behind the universe, Sayers avoided the problematic qualifications and quandaries manifest in postmodern attempts to authorize human agency.

The performative power of the word is upstaged by the Performing Word:

> . . . He to whom the whole is known,
> Being Himself the Architect
> The Craftsman and the Corner-stone. (42–44)

Believing in the ontological absolute of Trinitarian Love, Sayers was not traumatized by the epistemological ambiguities that mark human performances.[2]

This Writing Performance not excepted.

Notes

Introduction

1. I discovered that Susan Rowland reads the novel as I do, wherein "advertising becom[es] a metaphor for identity as a series of performing masks." See *From Agatha Christie to Ruth Rendell* 52.
2. The diversity of her performances became quite evident at a Sayers conference held in fall of 2000 at Grove City College, Pennsylvania, which drew not only members of a Peter Wimsey fan club, but scholars trained in theology, philosophy, medievalism, pedagogy, media studies, critical theory, dramaturgy, and literary criticism. Among the theologians and religious historians who have cited Sayers on the Trinity are Cyril Richardson (1958), Ted Peters (1993), and Linwood Urban (1995). The American pedagogical system inspired by Sayers's essay is discussed in *Recovering the Lost Tools of Learning* by Douglas Wilson.
3. Elin Diamond, *Performance* 2. Alan Read titled his book on performance theory *Theatre and Everyday Life: An Ethics of Performance.* For a helpful introduction to the development of performance studies see W. B. Worthen, "Drama, Performativity, and Performance."
4. Andrew Parker and Eve Kosofsky Sedgwick 2; Carlos Alonso, "Editor's Column: Position Paper" 204.
5. Richard Rorty, *Contingency, Irony, and Solidarity* 52. Hereafter cited in the text as (*CIS*).
6. Jacques Derrida, *Of Grammatology* 34–35, 158.
7. Roland Barthes, *Image-Music-Text* 143.
8. Judith Butler, *Bodies That Matter: On the Discursive Limits of "Sex"* 95.
9. Butler does assert the possibility of agency, but only through the spectacularizing of the codes which define "acceptable" human behavior: a parodic performance that can only be enacted, it would seem, by socially "unacceptable" people who feel marginalized by those codes and hence can sense that they are fictions which regulate behavior.
10. For Sayers, "the idea that the heavens decide man's destiny" is "rigid materialist determinism—repugnant to the Church, because it denies freewill" (*Ltrs* 4. 399).

11. Jacques Barzun describes Sayers with words that could easily refer to Rorty: "explicitly a pragmatic relativist" (*From Dawn to Decadence* 743).
12. Ludwig Wittgenstein, *Tractatus Logico-philosophicus* 151.
13. This parallel between the creations of God and of humans she will elaborate years later in *The Mind of the Maker*.
14. Catherine Kenney, *The Remarkable Case of Dorothy L. Sayers* xi.
15. Quoted in Chris Baldick, *The Social Mission of English Criticism: 1848–1932* 97. Baldick also quotes one of the Newbolt committee members, George Sampson, who wrote that the study of English literature "is not a routine but a religion . . . it is almost sacramental" (101).
16. I. A. Richards, *Science and Poetry* 95. Baldick documents the contempt of the Leavises for mass culture repeatedly in *The Social Mission of English Criticism* (157, 166, 184–85). In 1937, Q. D. Leavis wrote a nasty review of Sayers's last two novels for *Scrutiny*, describing Sayers's fiction as "stale, second-hand, hollow," her "deliberate indecency" as "odious," her "delving into emotional deeps" as "nauseating," her "moral burden" only such that "your frock MUST be well-cut." Leavis was especially hard on *Gaudy Night*, calling it a "combination of literary glibness and spiritual illiteracy," ultimately, a "menace to civilization" ("The Case of Miss Dorothy Sayers" 336–37, 338, 340).
17. Sandra M. Gilbert and Susan Gubar, *No Man's Land*, vol. 2, 355. In their other mention of Sayers, Gilbert and Gubar use Peter Wimsey as one among several examples of "wounded males who . . . appear in works by women" (2. 287).
18. Two of the most recent books on Sayers have focused on her detective fiction: Terrance L. Lewis, *Dorothy L. Sayers' Wimsey and Interwar British Society* (1994); Robert Kuhn McGregor, with Ethan Lewis, *Conundrums for the Long Week-End: England, Dorothy L. Sayers, and Lord Peter Wimsey* (2000).
19. Diamond 4.
20. Ibid.
21. Postmodernism, which celebrates pluralism, is itself pluralistic. Some critical theorists see it as a radical break from modernism's primary assumption: the emancipatory power of universal reason. Others regard the postmodern questioning of all cultural assumptions, including that of modernism, as merely extending the modernist commitment to emancipation. Richard Rorty, once identified with postmodern theory, now distinguishes himself from it; conversely, Fredric Jameson, a one-time denouncer of postmodernism, is now considered in alliance with it. One might argue, then, that postmodern discourse is as diverse as Christian discourse. Of course, it would be naive to assume that Sayers, though an outspoken Christian apologist, held all possible Christian beliefs (she claims to never have had a conversion experience, for example, and she was positively disdainful

of low church pietism). Similarly, it would be naive to assume that, because this book demonstrates parallels between her thought and certain postmodern constructs, it is arguing that she would assent to all paradigms considered "postmodern."

CHAPTER 1 THE PERFORMANCE BEGINS HER(E)

1. Dorothy L. Sayers and Jill Paton Walsh, *Thrones, Dominations* 97. Even though Walsh completed the novel, I have read Sayers's handwritten manuscript and can confirm that these are Sayers's own words. I thank the Marion E. Wade Center in Wheaton, Illinois, for making the manuscript available to me.
2. Robert Elbaz, *The Changing Nature of the Self* 155.
3. Sidonie Smith, "Construing Truth in Lying Mouths" 35.
4. The quotation is from Book XI, Chapter 17, of the *Confessions.* The translation is by Thomas C. Greene, who discusses how Augustine anticipates poststructuralist views of autobiography in a letter to the editor of *PMLA* ("Forum" 411).
5. *Ltrs* 1. 319. The next month she wrote Ivy, "I have for the moment abandoned my 'Memoirs,' as I shall have to get on with a new book" (*Ltrs* 1. 320). While some critics try to make a case that this "new book," *The Nine Tailors,* fulfilled Sayers's autobiographical goals, this still doesn't explain why Sayers replaced "My Edwardian Childhood" with *Cat o' Mary.*
6. Mikhail Bakhtin, *Author and Hero in Aesthetic Activity* 125–26.
7. Sayers, *The Nine Tailors* 238. She similarly mentions King Charles' head in a 1955 letter (Ltrs 4. 247).
8. As Foucault has argued, the very notion of an evolutionary continuity to identity is part of the modernist ideology of teleological progress that dictates the way people tell history, whether that of a nation or that of a self. Rather than following a linear progression, history, according to Foucault, is marked by "ruptures" in thought, causing discontinuous epochs which, like layers in an archeological dig, are not causally connected. See *The Archaeology of Knowledge.*
9. Bakhtin, *The Dialogic Imagination* 256.
10. Bakhtin, *Author and Hero* 15.
11. Smith 49.
12. This quotation appears on page 52 of *Cat.* However, the editor has substituted the word "exploration" for "reproduction." In the original autograph, Sayers has quite clearly written the word "reproduction."
13. James Brabazon, *Dorothy L. Sayers: A Biography* xviii; David Coomes, *Dorothy L. Sayers: A Careless Rage for Life* 26; Barbara Reynolds, *Dorothy L. Sayers: Her Life and Soul* 28. Hereafter, all will be cited in the text.

14. Sayers, *Wilkie Collins* 60. According to E. R. Gregory, Sayers was talking about publishing the *Life* as early as 1921 ("Wilkie Collins and Dorothy L. Sayers" 52). However, it would seem that the biography was at the forefront of her mind during the writing of *The Documents in the Case*, for she discusses Wilkie Collins in letters to Eustace Barton, her collaborator on *Documents* (Ltrs 1. 276–80).
15. Sayers, *Whose Body?* 122, 118. See also Sayers's short story "The Vindictive Story of the Footsteps That Ran," where Bunter says about a doctor, "he only sees the facts which fit in with the theory" (148).
16. Donald G. Marshall, "*Gaudy Night:* An Investigation of Truth" 100.
17. Mitzi Brunsdale, for example, has constructed a year-by-year chronology of Sayers's life in parallel with concurrent political and literary events (*Dorothy L. Sayers* 210–21).
18. Ralph Hone, *Dorothy L. Sayers: A Literary Biography* 45.
19. Hitchman, *Such a Strange Lady* 18.
20. Barbara Reynolds believes that Brabazon took the side of the producers because he himself was a television producer ("Reviews" 137).
21. In *Cat o' Mary*, Sayers writes, "Katherine's letters home always said that she was well and happy. . . . She would never have acknowledged that she was not a success" (91).
22. Only two months after Cournos left England in October of 1922, Sayers wrote to her parents about bringing Bill home to meet them; in April of 1923 she was pregnant.
23. In addition, "facts" often get blurred in the transmission. When I set out to find the "Ilford Lane" in Southbourne where, according to Reynolds, Sayers carried her baby to term (*Life* 124), I discovered that there is no "Ilford Lane" in Southbourne, but instead an "Iford Lane."
24. Smith 39.
25. *Ltrs* 4. 132–33, 16, 81. Sayers's distaste for biography must have been affected, as well, by worries over her secret son. Upon the death of Ivy Shrimpton, who raised John, along with other orphaned (and perhaps illegitimate) children, Sayers wrote, "I feel strongly that any letters or documents dealing with Ivy's various charges should be destroyed immediately. . . . Also, any old letters, etc. from me should also be burned" (*Ltrs* 4. 6).
26. Laura Marcus, "The Face of Autobiography" 19.
27. Ibid. 18.
28. Bakhtin, *Author and Hero* 166. For Bakhtin's definition of "heteroglossia," see *The Dialogic Imagination* 291–92.

CHAPTER 2 IDENTIFYING GENDER(ED) PERFORMANCES

1. James Brabazon, *Dorothy L. Sayers: A Biography* 122, 148; David Coomes, *Dorothy L. Sayers: A Careless Rage for Life* 9; Barbara

Reynolds, *Dorothy L. Sayers: Her Life and Soul* 230, 254–55, 175. In contrast, Valerie Pitt believes that "Sayers did not identify with Harriet Vane but with Wimsey," noting other critics who think the same ("Dorothy Sayers: The Predicaments of Women" 173–74). Hereafter, quotations from Reynolds' biography will be cited in the text.

2. See Brabazon 111; Coomes 37–38; and Janet Hitchman, *Such a Strange Lady* 43–44.
3. For the incident on the couch, see *The Devil is an English Gentleman*, vol. 2, 498–504. For lines that Cournos quotes from Sayers's letters, see pages 578–79 and 598 of *The Devil.*
4. Joan Riviere, "Womanliness as a Masquerade" 305.
5. The statement, written in a letter to Sayers's son after her death, is quoted in David Coomes's biography (221). In the same letter, Lewis also praised Sayers's "richly feminine qualities."
6. For references to these various costumes and charades, see vol. 1 of Sayers's letters: 18, 19, 33, 58–59, 26, 81, 83.
7. Susan Leonardi notes that for many Somerville women like Sayers, "College plays became an arena . . . for acting out self-sufficiency. The women could be both male and female" (*Dangerous by Degrees* 35).
8. This unpublished letter to Ivy Shrimpton, dated 6–11–08, can be found in File 47 of the Sayers holdings at The Marion E. Wade Center, Wheaton, IL.
9. John Thurmer describes Sayers as "more of a father than a mother" to John Anthony ("Sayers and Son" 76).
10. Judith Butler, like many performance theorists, argues otherwise in *Gender Trouble*, believing that "this construct called 'sex' is as culturally constructed as gender" (7).
11. Mary Eagleton, "Locating the Subject: Introduction" 342; Laurence Senelick, introduction to *Gender in Performance* ix.
12. I am of course describing the constructions of Western cultures. The fact that other societies write different scripts for male and female identity formation merely supports the constructedness of gender roles.
13. Pauline Adams, *Somerville for Women* 80. Sayers first presented "Are Women Human?" to a women's society in 1938; "The Human-Not-Quite-Human" was initially published in *Christendom: A Journal of Christian Sociology* (1941). In the body of my text, I will cite the essays by these dates, although the page numbers are taken from *Are Women Human?* published by InterVarsity Press. Both essays were reprinted by Sayers in *Unpopular Opinions* (1947).
14. Marvin Carlson, *Performance* 169.
15. Leonardi 52.
16. Mary Russo, "Female Grotesques" 224.
17. Alisa Solomon, *Re-dressing the Canon* 169.
18. This does not mean that discourse controls all human behavior, as some poststructuralists seem to imply. "Nothing is to be gained," notes Geraldine Harris, "if 'the' mimetic simply replaces 'the' real and

is perceived as a singular stable term that can operate as a universal touchstone for measuring the self-evident 'truth' of the fundamental fictionality and theatricality of all human experience" (*Staging Femininities* 75).

19. Barthes, *Image-Music-Text* 143.
20. This can also be said about the anti-Semitism of which she has been accused. Since she was ready to marry a Russian Jew, John Cournos, her "anti-Semitic" comments, like her male pronouns, are more likely a reflection of the anti-Semitism which inscribed her era: in other words, the staging of a writing performance. About the anti-Semitism of Sayers's day, see Gill Plain, *Women's Fiction of the Second World War* 9.
21. Bridget Elliot and Jo-Ann Wallace, *Women Artists and Writers* 153.
22. Qtd. in Susan Bordo, *Unbearable Weight* 241.
23. Deconstruction has been vilified as nihilistic, and in the hands of some practitioners it certainly seems to empty a text of any meaning. Derrida, however, meant something different: "I totally refuse the label of nihilism which has been ascribed to me and my American colleagues. Deconstruction is not an enclosure in nothingness, but an openness toward the other" ("Dialogue" 124). His openness toward the other has led him to theological interests—which may explain why he is cited today less by literary theorists than by scholars in religious studies. As Bruce W. Holsinger notes, Derrida "has been engaging the tradition of apophatic spirituality that includes Meister Eckhart and Pseudo-Dionysius the Areopagite" ("Medieval Studies" 1224).
24. Qtd. in Elaine Showalter, *A Literature of Their Own* 186, nt. 6.
25. Sayers, *Busman's Honeymoon* 284. Quotations, hereafter cited in the text, are from the novel, rather than the play, unless otherwise noted.
26. For a helpful discussion of the female body as the site of pre-discursive knowledge, see Kathryn Bond Stockton, *God Between Their Lips.*
27. Margaret Homans, *Bearing the Word* 13. Hereafter cited in the text.
28. Donna Perreault, "Woman as Other" 123.
29. We don't know when or how Sayers met Mac, but considering that her affairs with John Cournos and Bill White occurred during and after the writing of *Whose Body?*, it is highly unlikely that Mac was a part of her life at this time.
30. Reynolds, *Life* 200–01.
31. Catherine Kenney, *Remarkable Case* 149.
32. Gayle F. Wald makes a similar point in "Strong Poison: Love and the Novelistic in Dorothy Sayers" 108.
33. Susan Rowland, *From Agatha Christie to Ruth Rendell* 72.
34. SueEllen Campbell, "The Detective Heroine" 16–17; Margot Peters and Agate Nesaule Krouse, "Women and Crime" 147.
35. In 1946, George Orwell wrote "Since 1918 . . . a detective story not containing a murder has been a great rarity" (qtd. in *Ltrs* 3. 227, nt. 2). Leonardi notes that readers objected to the fact that *Gaudy*

Night did not "properly conform to the genre" due to its lack of a corpse (65).

36. Donald Marshall, "*Gaudy Night:* An Investigation of Truth" 113.
37. Laura Mulvey, *Visual and Other Pleasures* 19.
38. Carolyn G. Heilbrun, *Hamlet's Mother* 257. Marion Frank makes a similar point: "A key scene of the novel is when Harriet studies Lord Peter's face in the same manner as women are usually the object of men's vision" ("The Transformation of a Genre" 91).
39. Joyce Lannom Watts notes that "Sayers was more able to deal with what psychologists today [1981] call the maleness and femaleness in each of us" ("The Androgynous Aspects of Sayers" 4).
40. "Talboys" was written in 1942, but not published until 1972.
41. In her book on androgyny, Heilbrun quotes from both "Are Women Human?" and "The Human-Not-Quite-Human." Citing Heilbrun as well as other helpful studies on androgyny, Elizabeth Trembley argues for an androgynous, rather than feminist, Harriet Vane (' "Collaring the Other Fellow's Property" ' 91).
42. Glenwood Irons, introduction to *Feminism in Women's Detective Fiction* xii.
43. Alzina Stone Dale's excellent introduction to the play outlines how Sayers and Byrne divided up the responsibilities.
44. See *Honeymoon* 11, 14, 24, 25. In *The Stars Look Down* A. J. Cronin writes, "From Hilda, David got one view of Barras. The other, of course, came from Arthur.... But while Hilda's disclosure wore the taint of hatred, Arthur's rang out like an ecstasy" (147).
45. Showalter, *A Literature of Their Own* 150.
46. Bakhtin, *Problems of Dostoevsky's Poetics* 88. This text is Bakhtin's revision of his earlier *Problems of Dostoevsky's Creative Work.*
47. Bakhtin, *The Dialogic Imagination* 292.
48. Gill Plain makes the case that Bunter "assumes the nurturing role normally associated with the biological mother" (*Women's Fiction* 52).
49. Sayers, *Clouds of Witness* 162. Barbara Reynolds, like many others, aligns Cournos with Philip Boyes, Harriet's lover who is murdered in *Strong Poison* (*Life* 230). Brabazon notes, however, that *Clouds*, "the book she struggled with throughout this long, tense period, is concerned with sexual passion in a way that none of her other books is" (106). I would like to suggest that the marriage between Lady Mary and the middle-class Inspector Parker adumbrates the marriage between Lord Peter and Harriet Vane, a commoner, both women having been mistreated by their Russian-born lovers.
50. See note 8.
51. Parallels could also be made between Sayers and Miss Meteyard in *Murder Must Advertise* as well as Hilary Thorpe in *The Nine Tailors.* See "Sayers's Alter Egos" by Philip L. Scowcroft.
52. "Janet Reed" is the nom de plume of a character named Edith Daybrook. However, since Edith's creator preferred to be known as

"Dorothy L. Sayers"—the name that identified her with her work—rather than by her married name, Dorothy Fleming, I shall refer to Edith as Janet throughout my discussion.

53. *Love All* 191–92. Quotations will be cited in my text.
54. In 1994, Kathryn Bond Stockton employed the doctrine of *felix culpa* to explain the feminist desire to escape the figurative realm of discourse, explaining that "Now more than ever, post-structuralist feminists are losing their hope in 'positive' projects that directly deliver 'the' feminine difference and are placing hope, instead, in the failure of the dominant constructions that would seek to fix them as women" (*God Between Their Lips* 10). This "failure" of paternalistic language would be a "fortunate fall."

Chapter 3 The Performance Builds: Sayers's Architectural Imagination

1. About the pin see Brabazon, *Dorothy L. Sayers* 51. As to the stole, in February of 1919 Sayers wrote her mother about how unfashionable her "dear Percival" has become, which must mean she had acquired it several years previously (*Ltrs* 1. 149). Ralph Hone makes the mistake of calling Allen "Hugh Percival Allen" (*Poetry* 44). According to Cyril Bailey's biography, the name was always Hugh Percy Allen, though he was often addressed "H. P. A.," thus explaining the title of Sayers's poem, "To H. P. A." Over twenty years after Sayers left the choir, Allen attended and chatted with her about an Oxford staging of *The Zeal of Thy House* (Cyril Bailey, *Hugh Percy Allen* 159, 119–20).
2. See Hone's comments on page 44 of *Poetry of Dorothy L. Sayers.* All quotations of Sayers's poetry are taken from this anthology, unless otherwise noted. Line numbers follow each poem extract.
3. See the poems "The Last Castle," "Peredur," "Matter of Brittany," and "Lay" in *Poetry.*
4. See Hone's comment on page 78 of *Poetry.* Sayers's praise of the Mass in B Minor came in May 1914; by the next February she was to write, "I am now coping with the Doctor on another tack—he seems inclined to add me to a long procession of little tame cats who have adorned his organ-loft in succession. This is a nuisance, because I do not want to be anybody's tame cat. . . . I'll not dance to his piping! . . . Still, out of his own subject, I can meet him on an equality, and I will" (*Ltrs* 1. 107). We should not be surprised, then, when "From the Mass in B Minor" describes two knights meeting on an equality to fight: "I have met you foot to foot, I have fought you face to face, / I have held my own against you and lost no inch of place" (1–2). In the third stanza, Sayers writes, "Did you curse my insolence / And over-confidence?"—perhaps referring to her impersonation of him in the going-down play.
5. As late as 1956, the year before she died, Sayers was to write, "it is the business of the workers to serve the work" ("Sacred Plays" 24).

6. Sayers's poetic construction is made of words just as the bell in the poem is covered with words, anticipating the inscribed bells in *The Nine Tailors* almost twenty years later.
7. Sayers borrowed the image of the "lodge in a cucumber field" from the Old Testament: Isaiah 1: 8. In the introductions to her translations of Dante, Sayers repeatedly alludes to the lovely architecture of the poem (*Hell* 11; *Purgatory* 27, 53). Furthermore, when she died in 1957, she left unfinished a novel about Dante and his daughter, having first completed a floor-plan for as well as a description of the building that would house them.
8. Barbara Reynolds, "Fifty Years On: Dorothy L. Sayers and Dante" 6.
9. All of the quotations about Bluntisham Rectory can be found on pages 20–22 of "My Edwardian Childhood." One wonders how much Sayers's interest in architecture might have been influenced by a "solemn treat" enjoyed in Bluntisham: the trying on of her father's masonic regalia (EC 21). The masons, of course, tie the mysteries of their order to the heritage of the freemasons, builders of stone edifices.
10. Robert Harbison, *Eccentric Spaces* 73.
11. Brabazon 18–19.
12. Matthew Arnold, patron saint of the modernist religion of literature, "lavished . . ., extremes of adulation upon Oxford University," some of which were published in *Culture and Anarchy*. See Chris Baldick, *The Social Mission of English Criticism* 46–48.
13. T. E. Hulme, *Speculations* 135, 134; Walter Pater, *Appreciations* 18; Cleanth Brooks, *The Well Wrought Urn* 203; E. M. Forster, *Two Cheers for Democracy* 80, 82; Northrop Frye, *Anatomy of Criticism* 74. The Gosse quotation is from Baldick (229), who discusses the influence of both Gosse and Pater on Sayers's generation (134, 51). *The Verbal Icon* names a book by the American New Critic W. K. Wimsatt, Jr. Sayers was familiar with the New Criticism, which she alludes to in *The Poetry of Search and the Poetry of Statement* (14). Hans Frei summarizes the New Critical aesthetic with, "The literary text itself had an unchangeable, almost sacred, status conferred upon it and became a self-enclosed imagistic world, structured by such devices as paradox and irony, which the second-order commentator must, above all, leave as they are and not translate into some didactic 'meaning' by way of prose paraphrase" ("The 'Literal Reading' of Biblical Narrative" 44).
14. Wesley A. Kort spends a chapter of *C. S. Lewis Then and Now* discussing the importance of architectural imagery in Lewis's writing.
15. Edmund Wilson, *Axel's Castle* 133. Wilson ends *Axel's Castle* stating, "The writers with whom I have here been concerned have not only, then, given us works of literature which, for intensity, brilliance and boldness as well as for an *architectural* genius, an intellectual mastery of their materials, rare among their Romantic predecessors, are probably comparable to the work of any time" (235, emphasis mine). For

the influence of Coleridge on the "absolute structure" of poetry, see also Charles Altieri, *Enlarging the Temple*; Cleanth Brooks, *The Well Wrought Urn*; and W. K., Jr., Wimsatt, *The Verbal Icon*.

16. G. Wilson Knight, *The Starlit Dome* 56.
17. Sayers alludes to this line from "Kubla Khan" in her 1937 essay "Gaudy Night": "The American sales, however, contradicted this ancestral voice prophesying doom" (218). The quotations of "Kubla Khan" are from *The Complete Poetical Works*, ed. Ernest Hartley Coleridge.
18. The word "enchanted" is used by both Coleridge and Sayers. See "Kubla Khan," line 14, and "The Last Castle," line 23.
19. Sayers also referred to the devastation of the temporal in "Lay," worrying that Oxford might "crumble into the gulf of Time" (line 30). Later in life she was to read and recommend to others John Livingston Lowes's book on Coleridge, *The Road to Xanadu* (*Ltrs* 4. 148).
20. In his biography, Ralph Hone quotes an interview in which Sayers stated, "I had leanings towards poetry, but it didn't pay" (*Dorothy L. Sayers* 69).
21. A collection of Cournos's poems, *In Exile*, came out in 1923, the same year Sayers's first novel, *Whose Body?*, was published.
22. Forster is quoted by Sayers in "The Omnibus of Crime" 101; Edmund Wilson, "Who Cares Who Killed Roger Ackroyd?" 392, 394, 396. Catherine Kenney says that the tone of Wilson's essay seems "suitable, perhaps, for a new edition of the New England Primer or an eighteenth-century English tract against the evils of novel-reading" (*Remarkable Case* 58). Kenney's analogy, making it sound as though Wilson were an anachronistic crank, obscures the modernist dogmas Sayers was defying when she began writing detective fiction.
23. Sigurd Burckhardt, *Shakespearean Meanings* 29.
24. GN 208. Sayers's perception about detective fiction was endorsed 44 years later by Michael Holquist, who writes that the genre "was simply ignored by critics who felt it was beneath their dignity to engage" ("The Mystery of Detective Fiction" 186).
25. Marvin Carlson, *Performance* 173. The architecture of *Nine Tailors* seems to have been influenced by a church in John Meade Falkner's novel *The Nebuly Coat* (1903), about which Sayers once said, "the chief character in the story is the great church tower" (qtd. in Kenney 57–58).
26. I. A. Richards, *Practical Criticism* 320.
27. Sayers admits that *The Five Red Herrings*, her seventh novel, was a regression in her literary abilities (GN 209). Repeatedly in her reviews of detective fiction published in *The Sunday Times* Sayers uses machinery metaphors to indicate deficient writing performances. See Ralph Hone, "Dorothy L. Sayers: Critic of Detective Fiction" (46, 47, 49).
28. Nancy-Lou Patterson, "Beneath That Ancient Roof" 43. For another invective against machines, see the letter Sayers wrote to C. S. Lewis on December 8, 1945 (*Ltrs* 3. 180).

29. Eliade was most likely responding to Le Corbusier's famous aphorism: a house is "a machine for living in."
30. The bell described in Sayers's poem "To H. P. A." is also inscribed, as were all the bells at St. Mary's Church in Bluntisham, the bell tower having been refurbished by Sayers's father. Significantly, the newest bell in her father's church bore the name "Godfrey," the name that Sayers gives to the character in *Nine Tailors* who supervises the bells (24). Henry Sayers wrote "A Short Account of the Bells of St. Mary's Church" that can still be purchased in the Bluntisham church where he once served as rector.
31. Eliade 63.
32. Pater 14.
33. Repeatedly in her letters Sayers resists opportunities to be identified as a religious leader, saying that she wants neither to evangelize (2. 310) nor to exhort (2. 382). In contrast, she explicitly parallels herself with William of Sens (2. 47).
34. *Zeal* 27. Half a dozen years later, Cleanth Brooks was to use the phrase "poem in stone" to describe the Grecian Urn celebrated in Keats's famous poem (151).
35. In her introduction to *Performance and Cultural Politics*, Elin Diamond notes that "performance is always a doing and a thing done" (1).
36. *Zeal* 59–60. The coalescence of God's work with Christ's craft was not new for Sayers. In her second published book, *Catholic Tales and Christian Songs* (1918), a poem appears entitled "The Carpenter's Son," which presents the Master Architect God who transforms into Christ the carpenter. Its images were echoed a year later in a sixteen-stanza work entitled "The Journeyman," published in *The New Decameron* (1919).
37. Sayers did continue publishing short stories in the detective fiction genre, as well as collaborative works with the Detection Club. Nevertheless, detective fiction was no longer her primary focus. For an overview of Sayers's post-novelistic detective fiction, see Philip Scowcroft, "A Dorothy L. Sayers Crime Play Rediscovered."
38. Carlson 173.
39. Carlson quotes from Derrida's essay "The Ends of Man," inserting his own ellipses (173–74).
40. Qtd. in Carlson 174.

CHAPTER 4 MINDING THE PERFORMANCE: SAYERS'S LITERARY CRITICISM

1. *Zeal* 103. Michael's speech, which closes the published version of the play, was not a part of the original performance at Canterbury.
2. Sayers, "Statement of Aims" 279, emphasis mine. Sayers discusses the project in *Ltrs* 2. 137–38.

3. See *Ltrs* 3. 304, nt. 2, and 292, nt. 2. For overviews of the origins of *The Mind of the Maker* and its relation to Sayers's craft and theology, see Barbara Reynolds, *Dorothy L. Sayers*, chapter 23; Catherine Kenney, *The Remarkable Case*, chapter 13; and Margaret P. Hannay, ed., *As Her Whimsey Took Her*, section IV.
4. Citing Sayers in their discussions of the trinity are Ted Peters, *God as Trinity*, and Linwood Urban, *A Short History of Christian Thought.* Canon John Thurmer has written an entire book about the theological insight of *The Mind of the Maker*, calling it *A Detection of the Trinity.* Reformed scholars tend to be more suspicious of Sayers's trinitarian aesthetic. Roger Lundin dismisses her theory as demonstrating a "reactionary virus" among Christians who embrace outmoded paradigms once inimical to Christianity (*The Culture of Interpretation* 186) and Nicholas Wolterstorff takes Sayers to task in "Evangelicalism and the Arts."
5. Mikhail Bakhtin, *Author and Hero in Aesthetic Activity* 39. Hereafter cited in the text as (*Author*).
6. We don't know the exact details of what Bakhtin believed, primarily because of the Stalinist suppression of Christianity during much of his lifetime. Most scholars assume, and justify with biographical and literary data, a Russian Orthodox grounding for his faith. For helpful discussions of Bakhtin's religious thought, see Ruth Coates, *Christianity in Bakhtin*; Caryl Emerson, "Russian Orthodoxy and the Early Bakhtin"; and *Bakhtin and Religion*, edited by Susan M. Felch and Paul J. Contino.
7. David P. Haney, *Wordsworth and the Hermeneutics of Incarnation* 29.
8. Qtd. in James Brabazon, *Dorothy L. Sayers* 120.
9. Coates 87; Wolterstorff 466, 467.
10. Clement Greenberg, "Avant-Garde and Kitsch" 8.
11. Coates 64.
12. See W. K. Wimsatt Jr., *The Verbal Icon* 21–39.
13. Sayers credits *Principles of Art* for much of what she says in her 1944 address "Towards a Christian Aesthetic" (*Whimsical* 73), and she refers to Collingwood in her 1956 address "The Poetry of Search and the Poetry of Statement" (*Search* 15).
14. Wolterstorff 464, 465.
15. *Mind* 187, 204, 214, 122. Wolterstorff offers other legitimate criticisms of *Mind.* To make his point that Sayers does not acknowledge how "art participates in our fallenness" (467), he would have done well to quote and comment upon Sayers's problematic idea that "no creative Idea can be wholly destructive," for "the destruction is redeemed by its creative elements" (*Mind* 112).
16. Terry Eagleton, *Literary Theory* 74. Though no one before me has aligned Sayers with Reception Theory, I am not the first to suggest its resonances with Bakhtin. See Robert Stam, *Subversive Pleasures* 20–21; and David Shepherd, "Bakhtin and the Reader."

17. See Wolfgang Iser, *The Act of Reading* 73; and Hans Robert Jauss, *Toward an Aesthetic of Reception* 21, 29. Hereafter cited in text under the authors' names.
18. This discussion of Sayers, Collingwood, and Reception Theory first appeared in my essay "After the End of the Monster."
19. Poststructuralist "intertextuality" was largely influenced by Bakhtin. See Graham Allen, *Intertextuality*.
20. Mikhail Bakhtin, *The Dialogic Imagination* 156. Hereafter cited in the text as (*Dialogic*).
21. George N. Dove, *The Reader and the Detective Story* 63.
22. Bakhtin, *Problems of Dostoevsky's Poetics* 63.
23. Sayers, GN 209. Trevor Hall suggests that the format of *Documents* was inspired by Wilkie Collins's *The Moonstone* as well as E. C. Bentley's *Trent's Last Case* (*Dorothy L. Sayers* 63).
24. Bakhtin, *Problems* 55.
25. Thomas Michael Stein, "University Detective Fiction then and Now" 32.
26. *Busman's Honeymoon* is an exception, for reasons addressed in chapter 2.
27. Catherine Mowry LaCugna, "God in Communion with Us" 86.
28. David Lyle Jeffrey, *People of the Book* 186.
29. Derek Attridge, "Innovation, Literature, Ethics" 26, 25. Hereafter cited in the text.
30. As Robert Stam notes, Bakhtin's later essays on "Speech Genres" also refer to "the 'active influence of the reader,' whose 'apperceptive background of responsive understanding' must be taken into account" (*Subversive* 21).
31. This echoes Wyclif, for whom the "projected 'world' of the text . . . creates in the reader a *new being* as well as a 'new work' " (Jeffrey 186).
32. See her essay "The Divine Poet and the Angelic Doctor" (*Further* 38–52).
33. *Further* 197, 187. As though in echo of Sayers, Attridge cites Derrida's essay "Psyche: Invention of the Other" to argue for the doubleness of the phrase "the creation of the other." On the one hand, it can mean "bring[ing] into existence by skillful and imaginative intellectual labor an entity that is absolutely different from what is already in being." On the other hand, "The coming into being of the wholly new requires some relinquishment of intellectual control, and *the other* is a possible name for that to which control is ceded" (Attridge 21).

CHAPTER 5 THE PERFORMING WORD: SAYERS'S UNORTHODOX ORTHODOXY

1. She told one correspondent, "I never, so help me God, wanted to get entangled in religious apologetics" (*Ltrs* 4. 139).

2. For the tension between Modernists and Fundamentalists, see George Marsden, *Understanding Fundamentalism* 56–61.
3. Introduction to *King* 3.
4. Kathleen Nott, *The Emperor's Clothes* 31. Quotations, hereafter cited in the text, are from the 1958 edition. For an overview of the Kathleen Nott controversy see Volumes 2 and 3 of *VII: An Anglo-American Literary Review*, which not only published Sayers's 1954 rebuttal to Nott, but also invited Nott to respond to Sayers's comments for the 1982 issue. I similarly contrast Nott with Sayers and Rorty in my essay "Imbricating Faith and Learning."
5. Rudolph Bultmann, "New Testament and Mythology" 4.
6. Richard Rorty, *Contingency, Irony, and Solidarity* 52. Hereafter cited in the text as (*CIS*).
7. Kathleen Nott, "A Debate Deferred: Notes Towards a Reply" 48.
8. Sayers, "The Religions Behind the Nation" 68.
9. Rorty is quoting Donald Davidson (*CIS* 50).
10. Sayers was to repeat these "pagan" gestures years later. In the eleventh play of *The Man Born to Be King*, Pilate's wife dreams that a voice calls out "Great Pan is dead" at the moment of Christ's crucifixion; and in her essay "The Greatest Drama Ever Staged" Sayers parallels the resurrection of Osiris with that of Christ. All quotations of *Catholic Tales and Christian Songs* come from the original 1918 edition, and page numbers are cited in my text.
11. Sayers, introduction to *King* 6.
12. Sayers, "Creative Mind," *Whimsical* 109.
13. Throughout her writings during the war, Sayers uses the Nazis to exemplify "sheer barbarism" ("Religions Behind the Nation" 69). This will be addressed more thoroughly in the next chapter.
14. "Creative Mind," *Whimsical* 99.
15. Sayers, "A Debate Deferred: The Dogma in the Manger" 44.
16. Janice Brown, *The Seven Deadly Sins* 302.
17. John Thurmer, "Response to Sayers's 'Worship in the Anglican Church'" 57, nt. 2.
18. See, e.g., her essay "Creed or Chaos?" (*Creed* 27–54) and her letter to Father Herbert Kelly (*Ltrs* 2. 51–56).
19. Dale Irvin, *Christian Histories, Christian Traditioning* 29.
20. The proceedings of this address, which Sayers titled "The Church's Responsibility," were published in 1942. See 57–58.
21. Catherine Pickstock, "Radical Orthodoxy and the Mediation of Time" 67. While employing postmodern constructs, Radical Orthodoxy also foregrounds the limitations of postmodernism.
22. Laurence Paul Hemming, "Nihilism" 92.
23. James W. Welch, foreword to *King* 15. Sayers's interaction with the B.B.C. staff over this project, and the media-hyped scandal, has already been documented by James Brabazon, Barbara Reynolds, David

Coomes, and, most recently, Martha Eads. Once the plays were broadcast, positive response to their inspirational content far outweighed the negative.

24. Qtd. in David Coomes, *Dorothy L. Sayers* 18. Ironically, the most vociferous protests to *The Man Born to Be King* seemed to come from those who had not heard the radio plays. James Welch, in his foreword to the English edition of *King*, repeatedly cites the fact that protesters had no acquaintance with the plays.
25. Unpublished letter to B. E. Payne, February 11, 1942 (Marion E. Wade Center, File 520).
26. Unpublished letter to D. Burton, February 19, 1942 (Marion E. Wade Center, File 520).
27. Jeffrey, *People of the Book* xviii.
28. Daniel Williams, "Constantine, Nicaea and the 'Fall' of the Church" 122.
29. Sayers, "Debate" 37, 36. She also defines "dogma" as "opinion" in a 1940-letter to the editor of *The Spectator* (*Ltrs* 2. 170).
30. Joseph Huffman, "Faith, Reason, and the Text" 294. Huffman asserts that "postmodernism has in some significant respects returned to the Middle Ages for its deconstruction of modernist epistemology" 281.
31. Sayers refers to the "intellectual pattern" of Christianity in a 1943-letter to the Archbishop of Canterbury (*Ltrs* 2. 429).
32. Sayers, *Creed* 25. See also her comments to Father Kelly (*Ltrs* 2. 43–50).
33. Sayers, "Worship in the Anglican Church" 42. Hereafter cited in the text as (Worship). Colin Buchanan, a bishop in the Anglican Church, calls into question Sayers's assertion that individuals represent the whole church as they partake of the Eucharist. See his "Anglican Response" to Sayers's essay.
34. Frederick Christian Bauerschmidt, "Aesthetics: The Theological Sublime" 214; Graham Ward, "Bodies: The Displaced Body of Jesus Christ" 163.
35. Sayers, "Church's Responsibility" 59, 60.
36. John McGowan, *Postmodernism and Its Critics* 210.
37. James Brabazon, *Dorothy L. Sayers* 265. Chapter 1 of *Writing Performances* discusses how Brabazon's biography plots Sayers's life according to the conventions of Tragedy.
38. Qtd. in Barbara Reynolds, *The Passionate Intellect* 97.
39. In order to get around this metaphysical moment, Rorty admits that the word "cruelty" differs among interpretive communities. However, as noted by Stanley Fish, Rorty's belief in the "creative redescription" of the strong poet relies on faith as much as theism does. See J. Judd Owen's discussion of Fish and Rorty in *Religion and the Demise of Liberal Rationalism* 82, 139.

40. Sayers, "Religions Behind the Nation" 70, 75.
41. In *The Passionate Intellect* Barbara Reynolds discusses the influence upon Sayers of Dante's "Affirmation of Images" 172–74.
42. Qtd. in Dale Irvin 42.
43. Nott, "Debate" 41, 48.
44. Huffman 298.

Chapter 6 Begin Here: For the End(s) of the Performance

1. This description comes from Sayers's preface to the American edition of *Begin Here: A War-Time Essay* (v). The book was commissioned by Sayers's publisher, Victor Gollancz, who "asked her to write a Christmas message for the nation" (Reynolds, *Dorothy L. Sayers* 295). Sayers scholars, however, make only glancing reference to *Begin Here*. Though James Brabazon acknowledges its "profundity" and "flashes of insight," he includes Sayers's admission of its "errors and faults" (178) rather than the praise that she received from the Liberal M. P. Sir Richard Acland, himself an author and social commentator (see *Ltrs* 2. 160). Janice Brown gives the work slightly more attention in *The Seven Deadly Sins in the Work of Dorothy L. Sayers*, primarily because Sayers explicitly mentions gluttony, lust, and sloth in her text. However, like Catherine Kenney and Barbara Reynolds, Brown seems to regard Sayers's "War-Time Essay" as a mere dress-rehearsal for the more impressive performance of *The Mind of the Maker* soon to follow. Quotations from *Begin Here* are cited in the text as (*Begin*).
2. Catherine Kenney, *Remarkable Case* 249.
3. Judith Butler, *Bodies That Matter* 31.
4. Roland Barthes, *Image-Music-Text* 143.
5. Michel Foucault, *Power/Knowledge* 83. The discussion that follows draws upon the entry on Foucault I wrote for Kaufman and Baird, eds., *Twentieth Century Philosophy*.
6. Compare Marx's belief that "the exercise of human sense, powers and capacities is an absolute end in itself" (Terry Eagleton, *Literary Theory* 202) with Sayers's belief that when "the skill and labor of the worker are fully employed," they "acquire an absolute value" (*Creed* 72). Sayers, however, had no patience for Marxism as an ideology, criticizing both Fascism and Communism for their "suppression of organized bodies of opinion" (*Begin* 81).
7. In an essay written not long after *Begin Here*, Sayers acknowledges that "the word *dogma* is unpopular" and that the word "ideology" could be substituted for it (*Creed* 28), though in a letter around the same time, she argues against the notion that "ideology" is a more "liberal thing" than "dogma" (*Ltrs* 2. 157).
8. I. A. Richards, *Science and Poetry* 64.

9. Butler, *Bodies* 95.
10. Ibid. 9. Though she sounds very much like him, Butler doesn't wholeheartedly endorse Foucault, as he fails to account for "what is excluded from the economies of discursive intelligibility that he describes" (*Bodies* 35).
11. Critics distinguish between the "early" and the "late" Foucault, noting that the first two phases of his work discuss "technologies of domination" over mind and body, while the last phase explores "technologies of the self." In his later work, Foucault acknowledges the possibility of human agency—not through the discovery of self (which implies an essence independent of discourse) but through the reinvention of self. He advocates "micropolitics," where small groups of people contest varying manifestations of the discursive practices that constitute society. For a helpful discussion of this distinction, see Steven Best and Douglas Kellner, *Postmodern Theory* 34–75.
12. Geraldine Harris, *Staging Femininities* 14.
13. Elizabeth Abel, "Black Writing, White Reading" 478.
14. Margaret Homans, *Bearing the Word* 15.
15. Kathryn Bond Stockton, *God Between Their Lips* 8, 13–14.
16. Butler, *Bodies* 8.
17. As support for this point, note what Butler says about the materiality of the body: "To posit by way of language a materiality outside of language is still to posit that materiality, and the materiality so posited will retain that positing as its constitutive condition" (*Bodies* 30).
18. Qtd. in Harris 18. Stockton discusses the paradoxes that inscribe those poststructuralist feminists with whom Butler takes issue, such as Jane Gallop who situates the "body" as an "inscrutible given" (*God Between Their Lips* 17–19). Paradox seems unavoidable in efforts to establish agency.
19. Bradley Butterfield, "Ethical Value and Negative Aesthetics" 66.
20. See *Transfigurations: Theology and the French Feminists*, edited by Kim, St.Ville, and Simonaitis, and *Breaking Bread* by hooks and West.
21. See *Mind* 9, 10, 13, 17, 23, 26, 28, 36, 45, 183. The concept of "human nature" has been recuperated in the twenty-first century by evolutionary psychologists. M.I.T. psycholinguist Steven Pinker, e.g., defies leftist dogma by stating that the "existence of human nature is not a reactionary doctrine that dooms us to eternal oppression, violence, and greed" (qtd. in H. Allen Orr, "Darwinian Storytelling" 17). For an intelligent assessment of the "natural law" tradition and its relevance to postmodernity, see David L. Weeks, "The Uneasy Politics of Modern Evangelicalism."
22. I made these same points in my essay "Antiseptic Bakhtin," quoting physical anthropologist Christy Turner about his discoveries of cannibalism in the American Southwest: "From the standpoint of

sociobiology... cannibalism could well represent useful behavior done by well-adjusted, normal adults acting out their ultimate, evolutionarily channelled behavior" (Downing 23).

23. Stockton argues that "discourse on escapes from discourse" is "spiritual discourse" 10.
24. Jacques Derrida, "Structure, Sign, and Play" 85. Judith Butler, an inheritor of Derrida, makes the same point: "Against the claim that poststructuralism reduces all materiality to linguistic stuff, an argument is needed to show that to deconstruct matter is not to negate or do away with the usefulness of the term" (*Bodies* 30).
25. Laurence Paul Hemming, "Introduction" 7; Graham Ward, "Radical Orthodoxy and/as Cultural Politics" 104, 100. Radical Orthodoxy seeks to recuperate theology that precedes the "idolatrous" moves of Duns Scotus and William of Ockham in the late thirteenth century. See Phillip Blond, "Perception: From Modern Painting to the Vision in Christ" 232–34.
26. Laurence Paul Hemming, "*Quod Impossibile Est!* Aquinas and Radical Orthodoxy" 92, 93; John Milbank, *The Word Made Strange* 124, 126; and Oliver Davies, "Revelation and the Politics of Culture" 122–23. The word "participation" alludes to an idea Augustine borrowed from Plato: that created objects participate in Ideal qualities defined by and originating in the realm of the transcendent.
27. Sayers's desire to alter evil is explicitly stated in her "Statement of Aims" for *Bridgeheads*, the originators of which "believe that the only way to overcome evil is to transform it into a greater good"—something which cannot happen if "man sees himself as the helpless puppet of uncontrollable demonic forces" 279, 278.
28. Mikhail Bakhtin, *Author and Hero* 119.
29. Ibid. 144.
30. Sayers in a letter to her son, John Anthony, January 2, 1940, qtd. in Brabazon, *Dorothy L. Sayers* 163.

Encore: A Conclusion

1. Sayers, *Introductory Papers* 27.
2. Note Jacques Barzun's assessment of Sayers: "She was explicitly a pragmatic relativist" (*From Dawn to Decadence* 743).

Works Cited

Abel, Elizabeth. "Black Writing, White Reading: Race and the Politics of Feminist Interpretation." *Critical Inquiry* 19 (1993): 470–98.

Adams, Pauline. *Somerville for Women: An Oxford College 1879–1993.* Oxford: Oxford University Press, 1996.

Allen, Graham. *Intertextuality.* New York: Routledge, 2000.

Alonso, Carlos J. "Editor's Column: Position Paper." *PMLA* 117 (2002): 201–06.

Altieri, Charles. *Enlarging the Temple: New Directions in American Poetry During the 1960s.* London: Associate University Press, 1979.

Attridge, Derek. "Innovation, Literature, Ethics: Relating to the Other." *PMLA* 114 (1999): 20–31.

Bachelard, Gaston. *The Poetics of Space.* Translated by Maria Jolas. Boston: Beacon, 1969.

Bailey, Cyril. *Hugh Percy Allen.* Oxford: Oxford University Press, 1948.

Bakhtin, M. M. *Author and Hero in Aesthetic Activity.* In *Art and Answerability: Early Philosophical Essays by M. M. Bakhtin.* Translated by Vadim Liapunov, edited by Michael Holquist and Vadim Liapunov, 4–256. Austin: University of Texas Press, 1990.

———. *The Dialogic Imagination: Four Essays.* Translated by Caryl Emerson and Michael Holquist, edited by Michael Holquist. Austin: University of Texas Press, 1981.

———. *Problems of Dostoevsky's Poetics.* Translated by Caryl Emerson. Minneapolis: University of Minnesota Press, 1984.

———. *Toward a Philosophy of the Act.* Translated by Vadim Liapunov. Austin: University of Texas Press, 1993.

Baldick, Chris. *The Social Mission of English Criticism: 1848–1932.* Oxford: Clarendon, 1983.

Barthes, Roland. *Image-Music-Text.* Translated by Stephen Heath. London: Fontana, 1977.

Barzun, Jacques. *From Dawn to Decadence: 500 Years of Western Cultural Life—1500 to the Present.* New York: HarperCollins, 2000.

Bauerschmidt, Frederick Christian. "Aesthetics: The Theological Sublime." In *Radical Orthodoxy: A New Theology.* Edited by John Milbank, Catherine Pickstock, and Graham Ward, 201–19. New York: Routledge, 1999.

Benhabib, Seyla. *Situating the Self: Gender, Community and Postmodernism in Contemporary Ethics*. New York: Routledge, 1992.
Best, Steven and Douglas Kellner. *Postmodern Theory: Critical Interrogations*. New York: Guilford, 1991.
Blond, Phillip. "Perception: From Modern Painting to the Vision in Christ." In *Radical Orthodoxy*. Edited by John Milbank, Catherine Pickstock, and Graham Ward, 220–42. New York: Routledge, 1999.
Bordo, Susan. *Unbearable Weight: Feminism, Western Culture, and the Body*. Berkeley: University of California Press, 1993.
Brabazon, James. *Dorothy L. Sayers: A Biography*. New York: Scribner's, 1981.
Brombert, Victor. *The Romantic Prison: The French Tradition*. Princeton, NJ: Princeton University Press, 1978.
Brooks, Cleanth. *The Well Wrought Urn: Studies in the Structure of Poetry*. New York: Harcourt, Brace, 1947.
Brown, Janice. *The Seven Deadly Sins in the Work of Dorothy L. Sayers*. Kent, OH: Kent State University Press, 1998.
Brunsdale, Mitzi. *Dorothy L. Sayers: Solving the Mystery of Wickedness*. New York: Berg, 1990.
Buchanan, Colin. "Dorothy L. Sayers's 'Worship in the Anglican Church': An Anglican Response." *VII: An Anglo-American Literary Review* 13 (1996): 91–95.
Bultmann, Rudolph. "New Testament and Mythology: The Problem of Demythologizing the New Testament Proclamation." In *New Testament and Mythology and Other Basic Writings*. Edited and translated by Schubert M. Ogden, 1–44. Philadelphia: Fortress, 1984.
Burckhardt, Sigurd. *Shakespearean Meanings*. Princeton, NJ: Princeton University Press, 1968.
Butler, Judith. *Gender Trouble: Feminism and the Subversion of Identity*. New York: Routledge, 1990.
———. *Bodies That Matter: On the Discursive Limits of "Sex."* New York: Routledge, 1993.
Butterfield, Bradley. "Ethical Value and Negative Aesthetics: Reconsidering the Baudrillard-Ballard Connection." *PMLA* 114 (1999): 64–77.
Campbell, SueEllen. "The Detective Heroine and the Death of Her Hero: Dorothy Sayers to P. D. James." In *Feminism in Women's Detective Fiction*. Edited by Glenwood Irons, 12–28. Toronto: University of Toronto Press, 1995.
Carlson, Marvin. *Performance: A Critical Introduction*. New York: Routledge, 1996.
Chodorow, Nancy. *The Reproduction of Mothering: Psychoanalysis and the Sociology of Gender*. Berkeley: University of California Press, 1978.
Cixous, Hélène. "The Laugh of the Medusa." In *Critical Theory Since Plato*. Edited by Hazard Adams and LeRoy Searle, 309–20. Tallahassee: Florida State University Press, 1986.

Coates, Ruth. *Christianity in Bakhtin: God and the Exiled Author.* Cambridge: Cambridge University Press, 1998.

Coleridge, Samuel Taylor. "Kubla Khan." In *The Complete Poetical Works.* Edited by Ernest Hartley Coleridge, 2 vols. Oxford: Clarendon, 1912.

Collingwood, R. G. *The Principles of Art.* Oxford: Clarendon, 1938.

Coomes, David. *Dorothy L. Sayers: A Careless Rage for Life.* Batavia, IL: Lion, 1992.

Cournos, John. *The Devil Is An English Gentleman.* 2 vols. New York: Farrar & Rinehart, 1932.

Cronin, A. J. *The Stars Look Down.* Boston: Little, Brown, 1935.

Dale, Alzina Stone, ed. Introduction to *Love All: A Comedy of Manners by Dorothy L. Sayers together with Busman's Honeymoon: A Detective Comedy by Dorothy L. Sayers and Muriel St. Clare Byrne.* Kent, OH: Kent State University Press, 1984.

Davies, Oliver. "Revelation and the Politics of Culture: A Critical Assessment of the Theology of John Milbank." In *Radical Orthodoxy? A Catholic Enquiry.* Edited by Laurence Paul Hemming, 112–25. Burlington, VT: Ashgate, 2000.

Derrida, Jacques. "Dialogue with Jacques Derrida." In *Dialogues with Contemporary Continental Thinkers: The Phenomenological Heritage.* Edited by Richard Kearney. Manchester, U.K.: Manchester University Press, 1984.

———. *Of Grammatology.* Translated by Gayatri Chakravorty Spivak. Corrected edition. Baltimore, MD: Johns Hopkins University Press, 1998.

———. "Structure Sign and Play in the Discourse of the Human Sciences." In *Critical Theory Since 1965.* Edited by Hazard Adams and Leroy Searle, 83–94. Tallahassee: Florida State University Press, 1986.

Diamond, Elin, ed. Introduction to *Performance and Cultural Politics.* London: Routledge, 1996.

Dove, George N. *The Reader and the Detective Story.* Bowling Green, OH: Bowling Green State University Popular Press, 1997.

Downing, Crystal. "After the End of the Monster: Dorothy L. Sayers on Art." *Findings* (Winter 2002). <http://www.wilberforce.org/WilberforceForum/ChannelRoot/Features/FindingsOnline/After+the+End+of+the+Monster.htm>

———. "Antiseptic Bakhtin: A Dialogic Christian." *Pacific Coast Philology* 34 (1999): 18–31.

———. "Imbricating Faith and Learning: The Architectonics of Christian Scholarship." In *Scholarship and Christian Faith: Enlarging the Conversation.* Edited by Douglas Jacobsen and Rhonda H. Jacobsen. Oxford: Oxford University Press, 2004.

Eads, Martha Greene. "The Bible and the BBC: Dorothy L. Sayers's Working-Class Voices." *Cresset* 66 (November 2002): 6–10.

Eagleton, Mary. "Locating the Subject: Introduction." In *Feminist Literary Theory: A Reader.* Edited by Mary Eagleton, 339–50. Oxford: Blackwell, 1996.

Eagleton, Terry. *Literary Theory: An Introduction.* Oxford: Blackwell, 1983.

Elbaz, Robert. *The Changing Nature of the Self: A Critical Study of the Autobiographic Discourse.* Iowa City: University Iowa Press, 1987.

Eliade, Mircea. *The Sacred and the Profane: The Nature of Religion.* Translated by Willard R. Trask. New York: Harcourt, 1959.

Elliot, Bridget and Jo-Ann Wallace. *Women Artists and Writers: Modernist (Im)positionings.* New York: Routledge, 1994.

Emerson, Caryl. "Russian Orthodoxy and the Early Bakhtin." *Religion and Literature* 22 (1990): 109–31.

Felch, Susan M. and Paul J. Contino, eds. *Bakhtin and Religion: A Feeling for Faith.* Evanston, IL: Northwestern University Press, 2001.

Forster, E. M. *Two Cheers for Democracy.* New York: Harcourt, 1951.

Foucault, Michel. *The Archaeology of Knowledge and The Discourse on Language.* Translated by A. M. Sheridan Smith. New York: Pantheon, 1972.

———. *Power/Knowledge: Selected Interviews and Other Writings.* Translated by Colin Gordon. New York: Pantheon, 1980.

Frank, Marion. "The Transformation of a Genre – The Feminist Mystery Novel." In *Feminist Contributions to the Literary Canon: Setting Standards of Taste.* Edited by Susanne Fendler, 81–109. Lewiston, NY: Edwin Mellen, 1997.

Frei, Hans. "The 'Literal Reading' of Biblical Narrative in the Christian Tradition: Does It Stretch or Will It Break?" In *The Bible and the Narrative Tradition.* Edited by Frank McConnell, 36–75. New York: Oxford University Press, 1986.

Frye, Northrop. *Anatomy of Criticism: Four Essays.* Princeton, NJ: Princeton University Press, 1957.

Garber, Marjorie. *Academic Instincts.* Princeton, NJ: Princeton University Press, 2001.

Gilbert, Sandra M. and Susan Gubar. *The Madwoman in the Attic: The Woman Writer and the Nineteenth-Century Literary Imagination.* New Haven, CT: Yale University Press, 1975.

———. *No Man's Land: The Place of the Woman Writer in the Twentieth Century.* 3 vols. New Haven, CT: Yale University Press, 1994.

———. "Tradition and the Female Talent." In *The Poetics of Gender.* Edited by Nancy K. Miller, 183–207. New York: Columbia University Press, 1986.

Greenberg, Clement. "Avant-Garde and Kitsch." In *Perceptions and Judgments, 1939–1944.* Vol. 1 of *Clement Greenberg: The Collected Essays and Criticism.* Edited by John O'Brian, 5–22. Chicago: University of Chicago Press, 1986.

Greene, Thomas C. "Forum." *PMLA* 110 (1995): 411.

Gregory, E. R. "Wilkie Collins and Dorothy L. Sayers." In *As Her Whimsey Took Her: Critical Essays on the Work of Dorothy L. Sayers.* Edited by Margaret P. Hannay, 51–66. Kent, OH: Kent State University Press, 1979.

Hall, Trevor H. *Dorothy L. Sayers: Nine Literary Studies.* Hamden, CT: Archon, 1980.
Haney, David P. *William Wordsworth and the Hermeneutics of Incarnation.* University Park: Pennsylvania State University Press, 1993.
Hannay, Margaret P., ed. *As Her Whimsey Took Her: Critical Essays on the Work of Dorothy L. Sayers.* Kent, OH: Kent State University Press, 1979.
Harbison, Robert. *Eccentric Spaces.* New York: Knopf, 1977.
Harris, Geraldine. *Staging Femininities: Performance and Performativity.* New York: Manchester University Press, 1999.
Heilbrun, Carolyn G. *Hamlet's Mother and Other Women.* New York: Columbia University Press, 1990.
———. *Toward a Recognition of Androgyny.* New York: Knopf, 1973.
Hemming, Laurence Paul. "Introduction: Radical Orthodoxy's Appeal to Catholic Scholarship." In *Radical Orthodoxy? A Catholic Enquiry.* Edited by Laurence Paul Hemming, 3–19. Burlington, VT: Ashgate, 2000.
———. "Nihilism: Heidegger and the Grounds of Redemption." In *Radical Orthodoxy: A New Theology.* Edited by John Milbank, Catherine Pickstock, and Graham Ward, 91–108. New York: Routledge, 1999.
———. "*Quod Impossibile Est!* Aquinas and Radical Orthodoxy." In *Radical Orthodoxy? A Catholic Enquiry.* Edited by Laurence Paul Hemming, 76–96. Burlington, VT: Ashgate, 2000.
Hitchman, Janet. *Such a Strange Lady.* New York: Harper & Row, 1975.
Holquist, Michael. "The Mystery of Detective Fiction." *Meanjin* 40 (1981): 186–92.
Holsinger, Bruce W. "Medieval Studies, Postcolonial Studies, and the Genealogies of Critique." *Speculum* 77 (2002): 1195–1227.
Homans, Margaret. *Bearing the Word: Language and Female Experience in Nineteenth Century Women's Writing.* Chicago: University of Chicago Press, 1993.
Hone, Ralph. *Dorothy L. Sayers: A Literary Biography.* Kent, OH: Kent State University Press, 1979.
———. "Dorothy L. Sayers: Critic of Detective Fiction." *VII: An Anglo-American Literary Review* 6 (1985): 45–72.
hooks, bell and Cornel West. *Breaking Bread: Insurgent Black Intellectual Life.* Boston: South End Press, 1991.
Huffman, Joseph P. "Faith, Reason, and the Text: The Return of the Middle Ages in Postmodern Scholarship." *Christian Scholar's Review* 29 (1999): 281–301.
Hulme, T. E. *Speculations: Essays on Humanism and the Philosophy of Art.* Edited by Herbert Read. London: Routledge and Kegan Paul, 1958.
Hutcheon, Linda. *The Politics of Postmodernism.* New York: Routledge, 1989.
Ingarden, Roman. "Phenomenological Aesthetics: An Attempt at Defining Its Range." *Journal of Aesthetics and Art Criticism* 33 (1975): 257–69.
Irigaray, Luce. *This Sex Which Is Not One.* Translated by Catherine Porter. Ithaca, NY: Cornell University Press, 1985.

Irons, Glenwood. "Introduction: Gender and Genre: The Woman Detective and the Diffusion of Generic Voices." In *Feminism in Women's Detective Fiction*. Edited by Glenwood Irons, viv–xxiv. Toronto: University of Toronto Press, 1995.

Irvin, Dale T. *Christian Histories, Christian Traditioning: Rendering Accounts*. Maryknoll, New York: Orbis, 1998.

Iser, Wolfgang. *The Act of Reading: A Theory of Aesthetic Response*. Baltimore, MD: Johns Hopkins University Press, 1978.

Jameson, Fredric. *The Prison House of Language*. Princeton, NJ: Princeton University Press, 1972.

Jauss, Hans Robert. *Toward an Aesthetic of Reception*. Translated by Timothy Bahti. Minneapolis: University of Minnesota Press, 1982.

Jeffrey, David Lyle. *People of the Book: Christian Identity and Literary Culture*. Grand Rapids, MI: Eerdmans, 1996.

Kaufman, Walter and Forrest Baird, eds. *Philosophic Classics: The Twentieth Century*. Vol V. Englewood Cliffs, NJ: Prentice Hall, 1998.

Kenney, Catherine. *The Remarkable Case of Dorothy L. Sayers*. Kent, OH: Kent State University Press, 1990.

Kim, C. W. Maggie, Susan M. St. Ville, and Susan M. Simonaitis, eds. *Transfigurations: Theology and the French Feminists*. Minneapolis: Fortress, 1993.

Klein, Kathleen Gregory. *The Woman Detective: Gender and Genre*, 2nd ed. Chicago: University of Illinois Press, 1995.

Knight, G. Wilson. *The Starlit Dome: Studies in the Poetry of Vision*. London: Oxford University Press, 1941.

Kort, Wesley A. *C. S. Lewis Then and Now*. Oxford: Oxford University Press, 2001.

Kristeva, Julia. *Desire in Language: A Semiotic Approach to Literature and Art*. Translated by Thomas Gora, Alice Jardine, Leon S. Roudiez and edited by Reon S. Roudiez. New York: Columbia University Press, 1980.

Kuhn, Thomas S. *The Structure of Scientific Revolutions*. Chicago: University of Chicago Press, 1970.

Lacan, Jacques. *Ecrits*. Translated by Alan Sheridan. New York: Norton, 1977.

LaCugna, Catherine Mowry. "God in Communion with Us: The Trinity." In *Freeing Theology: The Essentials of Theology in Feminist Perspective*. Edited by C. M. LaCugna, 83–114. San Francisco: Harper, 1993.

Leavis, Q. D. "The Case of Miss Dorothy Sayers." Review of *Gaudy Night* and *Busman's Honeymoon*, by Dorothy L. Sayers. *Scrutiny* 6 (1937): 334–40.

Lewis, Terrance L. *Dorothy L. Sayers' Wimsey and Interwar British Society*. Lewiston, New York: Edwin Mellen, 1994.

Leonardi, Susan J. *Dangerous by Degrees: Women at Oxford and the Somerville College Novelists*. New Brunswick, NJ: Rutgers University Press, 1989.

Loeffelholz, Mary. *Experimental Lives: Women and Literature 1900–1945*. New York: Twaine, 1992.

Lundin, Roger. *The Culture of Interpretation: Christian Faith and the Postmodern World.* Grand Rapids, MI: Eerdmans, 1993.

Lyotard, Jean-Francois. *The Postmodern Condition: A Report on Knowledge.* Translated by Geoff Bennington and Brian Massumi. Minneapolis: University of Minnesota Press, 1984.

Marcus, Laura. "The Face of Autobiography." In *The Uses of Autobiography.* Edited by Julia Swindells, 13–23. London: Taylor & Frances, 1995.

Marsden, George. *Understanding Fundamentalism and Evangelicalism.* Grand Rapids, MI: Eerdmans, 1991.

Marshall, Donald G. "*Gaudy Night:* An Investigation of Truth." *VII: An Anglo-American Literary Review* 4 (1983): 98–114.

McGowan, John. *Postmodernism and Its Critics.* Ithaca, NY: Cornell University Press, 1991.

McGregor, Robert Kuhn, with Ethan Lewis. *Conundrums for the Long Week-End: England, Dorothy L. Sayers, and Lord Peter Wimsey.* Kent, OH: Kent State University Press, 2000.

Milbank, John. *The Word Made Strange: Theology, Language, Culture.* Oxford: Blackwell, 1997.

Mulvey, Laura. *Visual and Other Pleasures.* Bloomington, IN: Indiana University Press, 1989.

Niebuhr, Reinhold. *Beyond Tragedy: Essays on the Christian Interpretation of History.* New York: Scribner's, 1937.

Nott, Kathleen. *The Emperor's Clothes.* Bloomington, IN: Indiana University Press, 1958.

———. "A Debate Deferred: Notes Towards a Reply." *VII: An Anglo-American Literary Review* 3 (1982): 45–48.

Orr, H. Allen. "Darwinian Storytelling." Review of *The Blank Slate: The Modern Denial of Human Nature,* by Steven Pinker. *New York Review of Books,* February 27, 2003, 17–20.

Owen, J. Judd. *Religion and the Demise of Liberal Rationalism: The Foundational Crisis of the Separation of Church and State.* Chicago: University of Chicago Press, 2001.

Parker, Andrew and Eve Kosofsky Sedgwick, eds. Introduction to *Performativity and Performance.* New York: Routledge, 1995.

Pater, Walter. *Appreciations: With an Essay on Style.* New York: Macmillan, 1910.

Patterson, Nancy-Lou. "Beneath That Ancient Roof: The House as Symbol in Dorothy L. Sayers's Busman's Honeymoon." *Mythlore* 10 (Winter 1984): 39–46, 48.

Perreault, Donna. "Woman as Other, Other as Author, Author as . . . Man? The Aut*h*obiographical Dimension of *The Second Sex.*" In *Women and Autobiography.* Edited by Martine Watson Brownley and Allison B. Kimmich, 113–30. Wilmington, DE: Scholarly Resources, 1999.

Peters, Margot and Agate Nesaule Krouse. "Women and Crime: Sexism in Allingham, Sayers, and Christie." *Southwest Review* 59 (1974): 144–52.

Peters, Ted. *God as Trinity: Relationality and Temporality in Divine Life.* Louisville, KY: Westminster/John Knox, 1993.

Pickstock, Catherine. "Radical Orthodoxy and the Mediation of Time." In *Radical Orthodoxy? A Catholic Enquiry.* Edited by Laurence Paul Hemming, 63–75. Burlington, VT: Ashgate, 2000.

Pitt, Valerie. "Dorothy Sayers: The Predicaments of Women." *Literature and History* 14 (1988): 172–80.

Plain, Gill. *Women's Fiction of the Second World War: Gender, Power and Resistance.* New York: St. Martin's, 1996.

Ransom, John Crowe. *The New Criticism.* Norfolk, CT: New Directions, 1941.

Read, Alan. *Theatre and Everyday Life: An Ethics of Performance.* London: Routledge, 1993.

Reynolds, Barbara. *Dorothy L. Sayers: Her Life and Soul.* New York: St. Martin's, 1993.

———. "Fifty Years On: Dorothy L. Sayers and Dante." *VII: An Anglo-American Literary Review* 16 (1999): 3–6.

———. *The Passionate Intellect: Dorothy L. Sayers's Encounter with Dante.* Kent, OH: Kent State University Press, 1989.

———. "Reviews." *VII: An Anglo-American Literary Review* 2 (1981): 126–48.

Richards, I. A. *Practical Criticism: A Study of Literary Judgment.* London: Routledge & Kegan Paul, 1929.

———. *Science and Poetry.* New York: Norton, 1926.

Richardson, Cyril C. *The Doctrine of the Trinity.* New York: Abingdon, 1958.

Riviere, Joan. "Womanliness as a Masquerade." *International Journal of Psycho-analysis* 10 (1929): 303–17.

Rorty, Richard. *Contingency, Irony, and Solidarity.* Cambridge: Cambridge University Press, 1989.

———. *Philosophy and the Mirror of Nature.* Princeton, NJ: Princeton University Press, 1979.

Rowland, Susan. *From Agatha Christie to Ruth Rendell: British Women Writers in Detective and Crime Fiction.* New York: Palgrave, 2001.

Russo, Mary. "Female Grotesques: Carnival and Theory." In *Feminist Studies/Critical Studies.* Edited by Teresa de Lauretis, 213–29. Bloomington, IN: Indiana University Press, 1986.

Sayers, Dorothy L. *Begin Here: A War-Time Essay. 1940.* New York: Harcourt Brace, 1941.

———. *Busman's Honeymoon.* 1937. New York: HarperCollins, 1992.

———. *Catholic Tales and Christian Songs.* Oxford: Blackwell, 1918.

———. *Cat o' Mary.* In *Dorothy L. Sayers: Child and Woman of Her Time.* Edited by Barbara Reynolds, 25–160. Cambridge: Carole Green, 2002.

———. "The Church's Responsibility." In *The Life of the Church and the Order of Society Being the Proceedings of the Archbishop of York's Conference: Malvern 1941*, 57–78. London: Longmans, 1942.

———. *Clouds of Witness.* 1927. New York: HarperCollins, 1995.

———. *Creed or Chaos?* 1949. Manchester, NH: Sophia Institute, 1974.

———. "A Debate Deferred: The Dogma in the Manger." 1954. *VII: An Anglo-American Literary Review* 3 (1982): 35–44.

———. *The Devil to Pay*. 1939. In *Two Plays about God and Man*, 7–121. Noroton, CT: Vineyard, 1977.

———. *The Documents in the Case*. With Robert Eustace. 1930. New York: Harper and Row, 1987.

———. *The Emperor Constantine: A Chronicle*. 1951. Grand Rapids, MI: Eerdmans, 1976.

———. *The Five Red Herrings* (Retitled *Suspicious Characters*). San Francisco: Harper and Row, 1931.

———. *Further Papers on Dante*. London: Methuen, 1957.

———. *Gaudy Night*. 1935. New York: HarperCollins, 1995.

———. "Gaudy Night." 1937. In *The Art of the Mystery Story: A Collection of Critical Essays*. Edited by Howard Haycraft, 208–21. New York: Simon and Schuster, 1946.

———. *Have His Carcase*. 1932. In *On the Case With Lord Peter Wimsey: Three Complete Novels*. New York: Wings, 1991.

———. *He That Should Come*. 1939. In *Two Plays about God and Man*, 123–86. Noroton, CT: Vineyard, 1977.

———. Introduction to *Hell: The Comedy of Dante Alighieri the Florentine*. Translated by Dorothy L. Sayers. Harmondsworth, England: Penguin, 1949.

———. Introduction to *Purgatory: The Comedy of Dante Alighieri the Florentine*. Translated by Dorothy L. Sayers. Harmondsworth, England: Penguin, 1955.

———. Introduction to *A Time Is Born* by Garet Garrett. Oxford: Blackwell, 1945.

———. *Introductory Papers on Dante*. 1954. New York: Barnes and Noble, 1969.

———. "The Journeyman." In *The New Decameron: Volume the First*, 25–28. New York: McBride, 1919.

———. *The Just Vengeance*. 1946. In *Four Sacred Plays*, 275–352. London: Gollancz, 1948.

———. *The Letters of Dorothy L. Sayers*. Vol. 1, *1899 to 1936: The Making of a Detective Novelist*. Edited by Barbara Reynolds. New York: St. Martins, 1995.

———. *The Letters of Dorothy L. Sayers*. Vol. 2, *1937 to 1943: From Novelist to Playwright*. Edited by Barbara Reynolds. New York: St. Martins, 1997.

———. *The Letters of Dorothy L. Sayers*. Vol. 3, *1944 to 1950: A Noble Daring*. Edited by Barbara Reynolds. Cambridge: Dorothy L. Sayers Society, 1998.

———. *The Letters of Dorothy L. Sayers*. Vol. 4, *1951 to 1957: In the Midst of Life*. Edited by Barbara Reynolds. Cambridge: Dorothy L. Sayers Society, 2000.

Sayers, Dorothy L. "The Lost Tools of Learning." 1948. In *The Poetry of Search and the Poetry of Statement: and Other Posthumous Essays on Literature, Religion and Language*, 155–76. London: Gollancz, 1963.

———. *Love All.* In *Love All: A Comedy of Manners by Dorothy L. Sayers together with Busman's Honeymoon: A Detective Comedy by Dorothy L. Sayers and Muriel St. Clare Byrne.* Edited by Alzina Stone Dale. OH: Kent State University Press, 1984.

———. *The Man Born to be King: A Play-Cycle on the Life of Our Lord and Saviour Jesus Christ.* 1943. Grand Rapids, MI: Eerdmans, 1979.

———. *The Mind of the Maker.* 1941. San Francisco: Harper, 1979.

———. "My Edwardian Childhood." In *Dorothy L. Sayers: Child and Woman of Her Time.* Edited by Barbara Reynolds. 1–25. Cambridge: Carole Green, 2002.

———. *Murder Must Advertise.* 1933. New York: HarperCollins, 1995.

———. *The Nine Tailors.* 1934. New York: Harcourt Brace, 1962.

———. "The Omnibus of Crime." In *The Art of the Mystery Story: A Collection of Critical Essays.* Edited by Howard Haycraft, 71–109. New York: Simon and Schuster, 1946.

———. *OP. I.* Oxford: Blackwell, 1916.

———. Preface to *The Surprise*, by G. K. Chesterton. New York: Sheed and Ward, 1953.

———. *Poetry of Dorothy L. Sayers.* Edited by Ralph E. Hone. Cambridge: Dorothy L. Sayers Society, 1996.

———. *The Poetry of Search and the Poetry of Statement: and Other Posthumous Essays on Literature, Religion and Language.* London: Gollancz, 1963.

———. "The Religions Behind the Nation." In *The Church Looks Ahead: Broadcast Talks*, 67–78. London: Faber, 1941.

———. "Sacred Plays." *Episcopal Churchnews*, February 6, 1955, 24, 31–33.

———. "Statement of Aims for the proposed *Bridgehead* series of books." In *Dorothy L. Sayers: A Biography*, by James Brabazon, 278–82. New York: Scribner's, 1981.

———. *Strong Poison.* 1930. In *On the Case With Lord Peter Wimsey: Three Complete Novels.* New York: Wings, 1991.

———. "Talboys." In *Lord Peter: A Collection of All the Lord Peter Wimsey Stories.* Edited by James Sandoe, 431–53. New York: Harper and Row, 1972.

———. *Unnatural Death.* 1927. In *On the Case With Lord Peter Wimsey: Three Complete Novels.* New York: Wings, 1991.

———. *The Unpleasantness at the Bellona Club.* 1928. New York: HarperCollins, 1995.

———. *Unpopular Opinions: Twenty-One Essays.* New York: Harcourt, Brace, 1947.

———. "The Vindictive Story of the Footsteps That Ran." In *Lord Peter: A Collection of All the Lord Peter Wimsey Stories.* Edited by James Sandoe, 137–53. New York: Harper and Row, 1972.

———. *The Whimsical Christian: 18 Essays by Dorothy L. Sayers.* New York: Macmillan, 1978.

———. *Whose Body?* 1923. New York: Avon, 1961.

———. *Wilkie Collins.* Edited by E. R. Gregory. Toledo, OH: Friends of the University of Toledo, 1977.

———. "Worship in the Anglican Church." *VII: An Anglo-American Literary Review* 12 (1995): 34–48.

———. *The Zeal of Thy House.* 1937. In *Four Sacred Plays,* 7–103. London: Gollancz, 1948.

Sayers, Dorothy L. and Muriel St. Clare Byrne. *Busman's Honeymoon.* 1937. In *Love All: A Comedy of Manners by Dorothy L. Sayers together with Busman's Honeymoon: A Detective Comedy.* Edited by Alzina Stone Dale. Kent, OH: Kent State University Press, 1984.

Sayers, Dorothy L. and Jill Paton Walsh. *Thrones, Dominations.* New York: St. Martin's, 1998.

Scowcroft, Philip L. "A Dorothy L. Sayers Crime Play Rediscovered." *VII: An Anglo-American Literary Review* 9 (1988): 83–86.

———. "Sayer's [*sic*] Alter Egos." In *Sidelights on Sayers* 17 (February 1986): 13–16.

Senelick, Laurence, ed. Introduction to *Gender in Performance: The Presentation of Difference in the Performing Arts.* Hanover, NH: University Press of New England, 1992.

Shepherd, David. "Bakhtin and the Reader." In *Bakhtin and Cultural Theory.* Edited by Ken Hirschkop and David Shepherd, 136–54. New York: Manchester University Press, 2001.

Showalter, Elaine. *A Literature of Their Own: British Women Novelists From Brontë to Lessing.* Princeton, NJ: Princeton University Press, 1977.

Smith, Sidonie. "Construing Truth in Lying Mouths: Truthtelling in Women's Autobiography." In *Women and Autobiography.* Edited by Martine Watson Brownley and Allison B. Kimmich, 33–52. Wilmington, DE: Scholarly Resources, 1999.

Solomon, Alisa. *Re-dressing the Canon: Essays on Theater and Gender.* New York: Routledge, 1997.

Stam, Robert. *Subversive Pleasures: Bakhtin, Cultural Criticism, and Film.* Baltimore, MD: Johns Hopkins University Press, 1989.

Stein, Thomas Michael. "University Detective Fiction Then and Now: Dorothy L. Sayers's *Gaudy Night* and Amanda Cross's *Death in a Tenured Position.*" *VII: An Anglo-American Literary Review* 10 (1993): 31–40.

Stockton, Kathryn Bond. *God Between Their Lips: Desire Between Women in Irigaragy, Brontë, and Eliot.* Stanford, CA: Stanford University Press, 1994.

Suleiman, Susan. "Introduction: Varieties of Audience-Oriented Criticism." In *The Reader in the Text.* Edited by Susan Suleiman, 3–45. Princeton, NJ: Princeton University Press, 1980.

Thurmer, John. *A Detection of the Trinity*. Exeter, UK: Paternoster, 1974.
———. "Response to Sayers's 'Worship in the Anglican Church.'" *VII: An Anglo-American Literary Review* 14 (1997): 53–58.
———. "Sayers and Son." *VII: An Anglo-American Literary Review* 16 (1999): 73–78.
Trembley, Elizabeth A. "'Collaring the Other Fellow's Property': Feminism Reads Dorothy L. Sayers." In *Women Times Three: Writers, Detectives, Readers*. Edited by Kathleen Gregory Klein, 81–99. Bowling Green, OH: Bowling Green State University Popular Press, 1995.
Urban, Linwood. *A Short History of Christian Thought*. Oxford: Oxford University Press, 1995.
Wald, Gayle F. "Strong Poison: Love and the Novelistic in Dorothy Sayers." In *The Cunning Craft: Original Essays on Detective Fiction and Contemporary Literary Theory*. Edited by Ronald G. Walker and June M. Frazer, 98–108. Malcomb, IL: Western Illinois University Press, 1990.
Ward, Graham. "Bodies: The Displaced Body of Jesus Christ." In *Radical Orthodoxy: A New Theology*. Edited by John Milbank, Catherine Pickstock, and Graham Ward, 163–81. New York: Routledge, 1999.
———. "Radical Orthodoxy and/as Cultural Politics." In *Radical Orthodoxy? A Catholic Enquiry*. Edited by Laurence Paul Hemming, 97–111. Burlington, VT: Ashgate, 2000.
Watts, Joyce Lannom. "The Androgynous Aspects of Sayers in Harriet and Peter." *The Sayers Review* 4 (January 1981): 1–11.
Weeks, David L. "The Uneasy Politics of Modern Evangelicalism." *Christian Scholar's Review* 30 (Summer 2001): 403–18.
Welch, James W. Foreword to *The Man Born to be King: A Play-Cycle on the Life of Our Lord and Saviour Jesus Christ*, by Dorothy L. Sayers, 9–16. London: Gollancz, 1946.
White, Hayden. *Metahistory: The Historical Imagination in Nineteenth-Century Europe*. Baltimore, MD: Johns Hopkins University Press, 1973.
Williams, Daniel H. "Constantine, Nicaea and the 'Fall' of the Church." In *Christian Origins: Theology, Rhetoric and Community*. Edited by Lewis Ayres and Gareth Jones, 117–136. New York: Routledge, 1998.
Wilson, Douglas. *Recovering the Lost Tools of Learning: An Approach to Distinctively Christian Education*. Wheaton, IL: Crossway, 1991.
Wilson, Edmund. *Axel's Castle: A Study in the Imaginative Literature of 1870–1930*. London: Fontana, 1984.
———. "Who Cares Who Killed Roger Ackroyd?" In *The Art of the Mystery Story: A Collection of Critical Essays*. Edited by Howard Haycraft, 390–97. New York: Simon and Schuster, 1946.
Wimsatt, W. K., Jr. *The Verbal Icon: Studies in the Meaning of Poetry*. New York: Farrar, Straus, 1964.
Wittgenstein, Ludwig. *Tractatus Logico-Philosophicus*. Translated by D. F. Pears and B. F. Guiness. New York: Humanities Press, 1961.

Wolterstorff, Nicholas. "Evangelicalism and the Arts." *Christian Scholar's Review* 17 (1988): 449–73.

———. *Reason Within the Bounds of Religion*, 2nd ed. Grand Rapids, MI: Eerdmans, 1984.

Worthen, W. B. "Drama, Performativity, and Performance." *PMLA* 113 (1998): 1093–1107.

Index

Works by Sayers (DLS) are listed by title; works by others appear under the author's name